CONSTRUCTION DESIGN ECONOMICS

D1742289

Duncan P. Cartlidge

HUTCHINSON OF LONDON

To Jacqueline and Amelia

Hutchinson & Co (Publishers) Ltd
3 Fitzroy Square, London W1

London Melbourne Sydney Auckland
Wellington Johannesburg and agencies
throughout the world

First published 1976
© Duncan P. Cartlidge 1976

Set in Monotype Times
Printed in Great Britain by The Anchor Press Ltd
and bound by Wm Brendon & Son Ltd
both of Tiptree, Essex

ISBN 0 09 126031 0

Contents

Preface

Although this book has been written by a quantity surveyor and building economist, the contents and the message that it contains are for the benefit of the entire building design team, i.e. project managers, architects, building cost consultants, engineers and builders themselves. When designing buildings the criteria that should be foremost in the minds of all the design team are the long term consequences and effects of specifying particular materials and components. After all, do we want to bequeath to our children buildings that are too costly to occupy, or even to demolish? This may seem fanciful, but it has already been predicted that a number of new office blocks in the City of London will be just this.

I would like to express my appreciation to my wife, Jacqueline, for her patience during the recent months and also for typing the final manuscript.

London   Duncan Cartlidge

1 Design Criteria
Part 1

Builders, property owners and members of the building design team are becoming increasingly more aware of the relationship that exists between the design of a building and its eventual overall cost. Consequently increasing pressure is being exerted on the design team to evaluate, before a project leaves the drawing board, the suitability from the cost and performance aspects of the design and the forms of construction and materials that it is intended to incorporate into a proposed project. Over the years it has been proved that the selection of the correct design, coupled with the correct choice of materials and systems, have long-term effects on the performance of a building. This evaluation usually involves an examination by the building economist of all the viable design alternatives at various stages in the design sequence. To do this it is necessary to know the expected performance and the overall costs of all the major components within a proposed project. The type of information that should be available before the evaluation of a particular component or element can begin is, for example:

(a) the design brief and user requirements
(b) the initial and long term costs associated with a component or element
(c) what maintenance costs, if any, are likely to be incurred and at what intervals
(d) the expected life of a component or element
(e) in the case of a component, the expected life of the structure into which it is to be incorporated. (It would be an unnecessary expense to build into a structure with an expected or planned short life, materials of a highly durable and maintenance-free nature.)

If short life buildings were cheap enough, a series of them might cost no more than the equivalent long life building and would appear to have the advantage that the design brief, that is the user requirements, might be revised every few years to keep the building up to date and the design working effectively. A recent report by a committee formed by the Department of the Environment found that there may be some economic justification for the use, not of short life buildings, but of short life finishes with a maintenance free life. These could be incorporated into a more durable structure allowing the internal arrangement, or the finishes of the building, to be altered during the expected life of the complete building. This principle is already applied to the design of shop fronts in fashionable town centres, where the façade must change almost as regularly as fashions do. With proposals like these it is not too difficult to see a time – indeed that time is almost upon us – when the design of a building will be like completing a giant jig-saw puzzle; the pieces of the puzzle will be building components, all with differing performances, life spans and maintenance problems. This prospect makes the increased use of design evaluation even more necessary in order to use compatible materials successfully within the same structure.

The money and manpower resources spent annually on maintenance are a substantial proportion of the country's total building activity. Furthermore, for maintenance, the output per man in money terms is much less than the equivalent for new work, with little chance of much increase in this output owing to the limited size and nature of maintenance work.

Present day interest rates, tax laws and shortage of capital often indicate that it is more economic to avoid the more expensive materials which require low maintenance and to select instead low cost components requiring high maintenance; but this indication takes no account of the inconvenience, disruption and loss of earning potential which arises with the need to repair or replace. There is no way of measuring this factor in national terms, or indeed in many individual cases, so it is, perhaps frequently ignored.

With the stock of buildings increasing in number year by year, the problem of future maintenance is continually growing, and the situation will become worse as the available labour finds more rewarding employment elsewhere.

The conflicting demands on the available capital resources to the nation or to any individual authority inevitably lead to the pruning of budgets. The result is that building cost limits are usually set at a level which leaves very little scope to allow a serious attempt to be made to reduce the problem which is being created for the future. There is a similar reluctance in the private sector to invest in higher initial costs in order to make savings in the future.

Another fact to consider is the uncertainty of the future, with changes due to company take-overs and mergers, and individuals moving more frequently from house to house, it is clear the unnecessary expenditure should be avoided unless it can be reclaimed in an increased value on sale.

A source of reference for such data as are required to evaluate a building design is the Building Maintenance Cost Information Service. This is a system for collecting and interpreting building maintenance and other property occupancy costs including cleaning and maintenance with relation to the expected life of particular components. The information is in a form that can be easily and meaningfully used by the building economist when comparing differing design solutions.

The importance of using materials and a form of construction that has been carefully evaluated can be illustrated by means of the following example. Since the war an alarming number of multi-storey structures have been constructed, both by local authorities and by private developers. At first one of the more popular, and it was thought less complicated forms of construction, was a reinforced concrete framed structure, with brickwork infill panels. On the face of it, a well-tried form of construction which offered no problems either with erection or maintenance. However, during the last few years numerous examples of severe cracking of the brickwork cladding have been reported. Cases have arisen in Leeds, one of the pioneers of high rise flats, Plymouth, Salford, Hull, Lewisham (London) and Clydebank. This means that the clients involved have now to pay considerably more than was originally envisaged for maintenance over the life of the building.

Many of the design characteristics of buildings are a direct outcome of design decision, or the quality of the construction. Part of the knowledge and skill demanded of the design team is the ability to take into account the continuing technical and economic consequences of design decisions and be able to evaluate the future operating costs of buildings.

When comparing the costs and the performance of various building design solutions, not only the initial costs are taken into account, but also where applicable any recurring costs due to maintenance, partial replacement, etc., that may reasonably be expected throughout the life of a building. Taking into account the effects of the costs of maintenance and so on may seem inconsequential, until it is realized that in 1969 the cost, in this country, of building maintenance was in excess of £2000 million.

Consideration, when designing, should be given to the way in which the completed building is to be disposed of. If the client is a developer the finished building will either be:

(a) sold or let for profit

(b) used by the client for his own use.

In the first instance the initial costs of any services etc. will be the responsibility of the client and will have to be paid for out of capital, which in turn will probably be borrowed, perhaps at a high rate of interest. If the completed building is sold or let then any subsequent running and maintenance costs will be paid for by the occupier and not the client. Therefore, in this case, finishes and construction systems with low initial costs but higher subsequent costs will be most beneficial to the client.

If, on the other hand, the client intends to use the building himself then obviously the maintenance costs will have to be paid for by him. Initial costs come out of capital and therefore are not eligible to be used as an allowance against income tax; however, maintenance costs can be used in this way and, therefore, are of benefit to the client in this respect.

In order to evaluate a design solution from the cost and performance aspects it is necessary to carry out a study of the constituent parts or elements of the building under consideration. Although these studies are carried out on an elemental basis (for a full list of recommended elements see Appendix A, page 85) and therefore each part of the building tends to be thought of in isolation, it should be remembered that many of the elements are functionally interdependent and the decision to change the construction of part of a building, for something apparently more beneficial, may in turn have an injurious effect on some other elements.

For example, it may be found that to substitute a cheaper form of material in the external walls of a building results in an increase in the heat loss, which in turn means that a larger and more powerful form of heating system will be required to maintain heat levels.

Almost certainly a larger system will cost more than the original system, both initially and in the long term, and may eventually outweigh any economic advantages that were thought to have been gained.

The economics associated with the design of major elements will now be considered, a major element being defined as a part of a building which generally contributes a high percentage to the total cost.

Substructure

The bearing capacity of the ground has been described as 'infinitely variable' and this comment can equally well apply to the design or the choice of foundation systems. It is this variability and uncertainty that makes this element so comparatively difficult to evaluate. There can be said to be four basic types of foundation systems, these are:
(a) strip foundations
(b) proprietary systems of foundations
(c) raft foundations
(d) piled foundations not included in (b).

The type of foundation system that will be used for a particular building will, to a large extent, be dictated by the physical characteristics of that building. Research and experience have proved that traditional strip foundations can be safely and economically used for buildings of up to four or even five storeys. At the other end of the scale the concept of multi-storey buildings would to a large extent be impossible if it were not for the use and the increased development of piled foundations, in all their forms. However, this is an over-generalization of the situation and there are many occasions when the relative economies of various types of foundation systems have to be examined, especially in the cases of low and medium rise buildings.

The type of foundation system that is probably used most often throughout the country is the traditional strip foundation. This system has the advantage of being easy to construct, generally requiring no special equipment; however it does suffer from the distinct disadvantage that it is labour intensive. An analysis of the costs that are involved with this element shows that probably more than any other element the majority of the total costs are attributable to expenses of labour. With the ever-increasing cost of employing labour in all its forms it is only logical that attempt should be made to reduce the percentage of labour costs within the total costs of the element. Figure 1 illustrates how labour costs have risen during recent years, compared to the increase that has taken place generally in the cost of materials. There is every indication that this trend will continue.

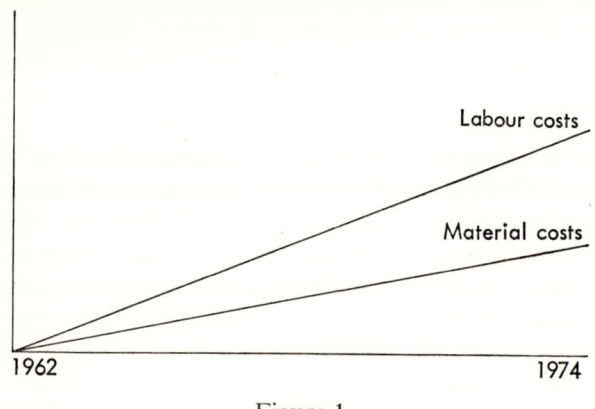

Figure 1

Traditional strip foundations are used mainly for low and medium rise buildings, where the bearing capacity of the ground presents no problems and where there are unlikely to be any heavy point loadings. However the rapid increase in the cost of labour has made the use of these foundations, which were for a long time considered to be an inexpensive design solution, an uneconomical proposition. According to *Building* the nationally negotiated wage claim in October 1972 had the effect of increasing the cost of employing labour by 26% and the general level of building prices by approximately 15%. Similarly the 1974 wage settlement was reflected in a rise in building prices generally during 1975.

If present trends continue and labour remains such an expensive commodity then the relationship that labour costs bear to the total cost, not just of this element, but of the whole building process, is a very significant factor in determining the price of buildings. As inevitable as the reduction in the amount of man hours for this element sounds, in order to achieve this it is necessary to introduce some form of industrialization and/or mechanization into the building process. In the early 1960s when various forms of industrialized building were being investigated *The Basingstoke Development Group* tried to produce a system of industrialized building for two storey houses. As with most systems of this kind it was found to be a relatively easy task to 'industrialize' the elements and the components of the work above ground level. However, when it came to the substructure it was found to be impossible to 'industrialize' any of the operations, leaving a total of 166 items that had to be performed either totally by hand or with the aid of mechanical plant.

Of the attempts that have been made to reduce the number of operations involving labour with this element, two noteworthy ones are the use of deep strip foundations, even on sites where the bearing capacity of the ground is adequate, and the so-called *Finchampstead Project*, which investigated the use of factory made components in the substructure. Both of these systems fall into the category of proprietary techniques and would therefore seem to suffer from serious financial set-backs which usually, in the final reckoning, make these inherently cheaper forms of construction only break even as far as costs are concerned. It appears that, especially in the case of the *Finchampstead Project*, the maximum economies are only to be obtained when it is possible to produce the components on which this system is based at a constant output over a reasonable period of time and when builders have become as familiar with new and developing techniques as with traditional forms of construction. The economies of industrialized building will be discussed in a later chapter.

Firstly, the use of deep strip foundations. The Cement and Concrete Association have, in the light of rising labour costs, recently promoted a system called 'trench fill' for constructing foundation systems for domestic dwellings. Figure 2 shows the comparative costs of trench fill compared to traditional strip foundations. It seems that the use of this type of foundation, which is material intensive, is an economic proposition in certain circumstances.

Reducing the amount of labour intensive operations required to construct a foundation system, in a manner described by the association, is claimed to produce savings of up to 33% compared with traditional strip foundations. A detailed comparison was carried out by S. Lazarus and Partners in association with Peter H. Hill and Partners and the subsequent report shows that on a typical block of five three storey houses there can be a saving of £52, equal

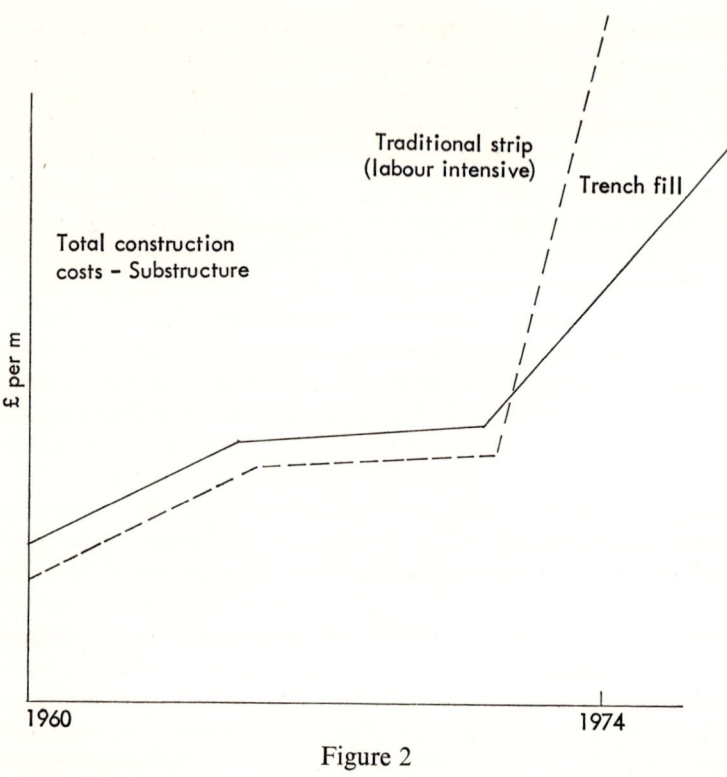

Figure 2

to approximately 6%, and on a typical block of two storey detached houses there was a total saving of £207, approximately 31%. In times of galloping inflation these figures at first glance appear to be outstanding; however, savings are cut if work cannot be carried out under near perfect conditions, that is on a dry, level site. But, conditions apart, this system has distinct advantages over labour intensive methods and is worthy of development.

The importance of developing an economic system of foundations for low and medium rise buildings lies in the fact that, as the latest housing and construction statistics show, only very few homes were approved to be built in high rise buildings. The construction of dwelling houses, both in public and private sectors, including associated works, account for approximately 35% of the total turn-over in the building industry, which is approximately equal to £1200 million in an average year.

The Finchampstead Project was a collaborative exercise between the Department of the Environment and the Building Research Station. A group was appointed to design a housing scheme for a site of five hectares in Finchampstead and it was decided that all the houses were to be terraced with a rectangular plan shape. The site was flat and presented no problems. Various methods were considered for providing a foundation system for the houses and despite the apparent simplicity of raft foundations the man hour requirements were high, about one third of the man hours required subsequently for the superstructure as a whole. Therefore the group set to work to design a more economical system which briefly involved casting *in situ* concrete foundation pads in a foundation trench, levelling these pads and then laying factory manufactured pre-cast concrete beams between the pads. It was possible to arrive at four beam lengths that in various combinations could make up all the lengths and depths required. The successful tender figure was slightly higher than the sum estimated for traditional foundations, but this, as has been stated earlier, may be due to the contractor's unfamiliarity with the method of construction. However, on site there were practically no subsequent complications and the final overall cost indicated a saving of 10% on the estimated cost of traditional strip foundations.

The effects of the cost of maintenance of foundations is seldom, if ever, a deciding factor in the choice of a particular system. This is due to the fact that a foundation's function is so basic that if designed and constructed properly, it should function adequately throughout the expected life of the building. Even with the comparatively untried methods as described above, it is doubtful whether there would be any maintenance costs to be taken into account and in any event to design a foundation system with a limited life span would indeed be a risky operation.

Raft foundations have rather special application and are used generally where no firm bearing strata exist at a reasonable depth and where loading is reasonably light and evenly distributed. This form of foundation may be the way to carry the load from lightly oaded columns to weak soil, but it suffers from the disadvantages associated with strip foundations, i.e. it is labour intensive. In fact, during the investigations to choose a type of foundation system for the *Finchampstead Project* it was discovered that although raft foundations may have provided a reasonably simple method of

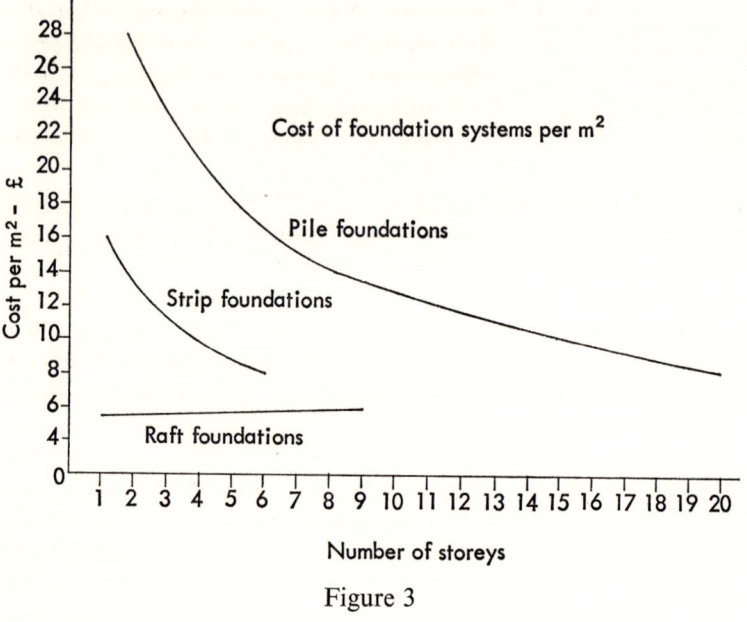

Figure 3

construction, about half the man hours required for the substructure as a whole were simply for laying services on, or under, the ground floor raft.

It is sometimes thought that piled foundations are necessarily an expensive solution to a foundation design problem and of course, on the face of it, a direct comparison of costs with most other foundation systems will show this to be correct (see figure 3). However, whenever piling in one of its many forms is used, it should offer an economically better solution than the other forms of foundation systems that have previously been discussed. Piled foundations are used principally in the construction of high rise buildings and the savings that can be achieved by building upwards instead of outwards and the total floor area it is possible to provide on one set of foundations, over which the cost is spread, make piling an economic device. One considerable advantage piled foundations have over other forms is the fact that construction need not be held up by periods of rainy or otherwise inclement weather.

Finally, the plan shape of the proposed building will have an important effect on the cost of this element, not only in the ways described in Chapter 6 of *Cost Planning and Building Economics*, but if the plan shape of a building develops from a square then the site costs involved in setting out the substructure also increase considerably. Recently a number of private developers and local authorities used multi-sided structures in an attempt to achieve the optimum design solution based on the facts that:

(a) the external walls of a circular plan shape will enclose the greatest amount of floor area of any plan shape but are expensive to construct in practice when working with curves

(b) the square is the most economical plan shape to use in practice.

Although exact figures are not available, the London Borough of Hillingdon have achieved a degree of success and produced an economic solution with a development of multi-sided shop units within the borough.

Therefore, the cost and design criteria to be taken into account when considering this element are:

(a) the physical characteristics of the superstructure, paying particular attention to the following elements: frame, upper floors and external walls, remembering that no element should be considered in isolation

(b) during present economic trends endeavour to minimize the amount of labour that is required

(c) consider the project to establish whether it is large enough or repetitive enough to benefit from the use of industrialized building techniques, for example, a housing scheme

(d) consider the best plan shape and the nature of the site

(e) finally try to establish if there are likely to be any maintenance costs during the expected life of the element.

The next major elements to be considered are:

Frame/Upper floors

Frame

Generally speaking, it is not necessary to use a frame for buildings with normal loadings up to four or even five storeys high, as it is possible to use load-bearing walls economically, together with traditional strip foundations described early in this chapter. However for taller buildings, where the total of the dead and superimposed loads becomes considerable, then it is usually not economical to use load-bearing external walls. They have to be so thick to withstand the loading that the usable floor area can be cut down considerably, and therefore it is better to use some kind of framework. The object of a frame is to transmit the loadings, via a series of beams and columns, to the foundations, thereby allowing the external walls to be comparatively thin.

This statement is, I admit, a sweeping generalization and there are, as with most things, exceptions to the rule, for example the World Trade Centre, New York, dubbed when it first opened 'the tallest building in the world'. It provides 10 million square feet of rentable office space in two, one hundred and ten storey, 1350 feet tower blocks. The blocks were designed so that 75% of the plan area is rentable space and yet this has been achieved by using a system of load-bearing external walls which completely eliminates the need for an internal system of columns. At the other end of the scale, steel framed two storey houses are being constructed on quite a large scale in the Midlands and the North of England. The system, its manufacturers contend, gives 'unlimited choice of design possibilities' and is economically viable due to ease and speed of erection on site.

The most commonly used types of frame are:

(a) *In situ* concrete
(b) Pre-cast concrete
(c) Structural steelwork

The case for concrete against steelwork

The regulations relating to building in this country stipulate that buildings intended to be used as hotels, offices, etc. should have good resistance to fire. Concrete has an inherent resistance to fire, whereas steelwork will buckle and twist when exposed to intense heat. Ironically it is often protected from fire damage by being coated with its own best competitor, concrete, although of course there are a score of other materials that are widely used for this purpose.

Steelwork therefore has the major advantage in that it is speedy to erect, much of the fabric can be carried out in 'shop conditions' and it therefore is unaffected by periods of inclement weather. In addition the initial cost of steelwork is less than its rivals, however, much of this advantage is removed when the cost of the necessary fire protection is added.

But not all buildings require such a high degree of fire resistance and in these areas steelwork has major cost advantages over its competitors. For example, it is ideally suited for use in single storey warehouses or factories, together with light weight cladding to the external walls.

In a project where speed is of the essence, as a general rule, precast concrete structural members are ideal. However, they are expensive and if it is decided to use them, the size of the members should be kept to a minimum. Also the site should not be so restricted so to hinder the delivery, unloading and transportation of, what could be, large pre-cast beams.

Finally, timber is coming more onto the scene as a structural material, its greatest disadvantage being its propensity to burn. However, quite recently the Bell Sports Centre in Perth was constructed using a laminated timber beam dome providing 3000 m² of floor area. Shortly before its opening it was engulfed for an hour by fire, even so the structure suffered little damage.

Timber is also an economical form of construction to use for the first floors of domestic dwellings. For larger and more complicated structures a number of reinforced concrete ribbed, trough and waffle floor sections have been developed with the object of reducing the volume of concrete required and therefore the cost. Pre-cast plate floors are available and are much more speedy to erect than the *in situ* systems described above. However, as we have seen before, pre-cast members are more expensive than the *in situ* ones.

Therefore, because each form of construction, used under ideal circumstances, offers saving in time and/or overall cost, the problem is deciding when to use what system and where, for this reason, it is best to evaluate each set of circumstances as they occur.

However, the size and arrangement of the columns can also be a critical determining factor when deciding on the type of frame. If the client is building for letting the efficiency of the design and especially the ratio of the gross to net lettable floor area is of utmost importance, because it is undesirable to have large areas of floor space taken up by columns. For example, in the City of London where office rents are very high, a tenant may argue that the areas taken up by columns should be deducted from the lettable floor area, hence a loss in revenue. In such a case it is more beneficial to have a few carefully positioned columns.

Like the previous element, substructure, there are very few long term maintenance costs associated with the frame or upper floors of a building, with the possible exception of repainting the exposed steelwork of a factory or warehouse every five years or so and therefore the initial cost of this element is likely to be the only cost. That is unless some sophisticated system like the one used for the new London Stock Exchange is specified, for it has to be estimated that it will cost as much to demolish the pre-stressed structure in eighty years' time, as it has to construct it, because of its complex 'umbrella' type structure.

Therefore the cost and design criteria relating to this element are likely to be:

(a) the functional requirements of the building
(b) the type of external walling or cladding to be used
(c) the order of priorities: for speed choose pre-cast, for economy *in situ* concrete
(d) does the structure lend itself to the use of a particular material, e.g. long, high sweeping concrete vaults and spans
(e) finally, the comments relating to labour efficiency and plan shape that were made in the element substructure, still apply here.

2 Design Criteria
Part 2

Roof

The RICS Standard Form of Cost Analysis requires that the cost of the roof element should be split down into the following:
(a) structure
(b) coverings
(c) drainage
After which it will be found that the first two parts, structure and coverings, will make up the majority of its elemental cost. The form of roof and its coverings will, to a great extent, be dictated by the use to which the building is to be put. I have, for this reason, attempted the dangerous practice of generally categorizing building types thus:

(i) Commercial and Domestic
In this category the choice is often between pitched or flat roof construction. A cost comparison between a timber pitched roof and a concrete flat roof, excluding the coverings, will show a concrete flat almost 35% more expensive than a pitched roof. However, savings as great as this may only be achieved with regular plan shapes; small insets or projections or other complications in the plan shape of a building with a timber pitched roof will considerably increase the amount of hips, valleys and intersections which will in turn increase the material, the labour content and the cost. Therefore, once again, the effect of the plan shape can be seen to be an important influence on costs.

The rule seems to be, for high rise buildings to have flat and not pitched roofs despite the savings outlined above. The main reasons are that pitched roofs on tall buildings look so incongruous and also that the roof area on tall buildings is used for plant rooms, tank rooms etc.

(ii) Industrial

Industrial buildings generally need roofs that provide large uninterrupted spans and also are required to incorporate some form of roof-lights. The cheapest solution in cases like this has been found to be an asbestos cement sandwich with translucent sheets on lightweight mild steel roof trusses. For buildings requiring exceptionally large spans or a higher degree of fire resistance there are a multitude of proprietary systems and materials some of which are capable of being erected very quickly, but at a higher cost than the system described above.

For flat roofs the choice of coverings seems to be sheet metal, asphalt or felt. Metal sheet roofing needs no regular treatment other than the removal of accumulated rubbish, and an inspection of joints and areas where the metals are stressed. Inspection of asphalt is advisable at intervals. The material is brittle when cold, and mechanical damage can occur. There may be blistering where there is no underfelt (for example, on steep slopes or up-standards), or signs of crack formation caused by structural movement, or general minor shrinkage causing movement around edges. All bitumen felt roofs should be inspected on a routine basis.

Regular treatment of bitumen felt is advisable only where the bitumen is exposed, but most such felts have mineral surfaces of some kind, which cannot be adequately treated, though minor local defects can be repaired by patching. Anything other than minor defects calls for replacement. Asphalt roofing properly designed and laid should prove capable of lasting 50–60 years; the natural ageing of bitumen felt is likely to limit its life to about 20 years.

Sheet metals may also, of course, be used for pitched roofs although this will prove to be expensive. The most economical and widely used type of covering for pitched roofs being concrete interlocking tiles on battens and felt. Metal sheet finishes are generally the most expensive, and minimum specification built up bitumen felts the cheapest. Asphalt may not be more expensive than the more sophisticated built up finishes or plastics. Natural rock asphalt is about one third more expensive than that with limestone aggregate. Costs vary with the complexity of the surfaces to be covered, and from this point of view, the design of roof surfaces should take careful account of the material to be used.

The durability of flat roof finishes is a major cause for concern, and it is advisable to consider at an early stage in design the recommendations of the relevant trade associations with regard to suitability for purpose within the cost target. Statements about the life expectations of roofing materials refer to perfect laying, which depends on the efficiency of site operations.

In calculations of cost, repairability is also important, particularly in the lower price ranges, where exceptionally long life is unlikely to be achieved. Fundamentally, only a well-built, well-specified roof is likely to prove economic in service.

What is the future for plastics in roofing?

The plastics industry now has probably the greatest opportunity in its history to make a significant impact on the roofing market. With the daily increasing world shortage of fuel, the need to conserve energy is becoming of paramount importance. This will inevitably lead to the introduction of higher standards for insulation in buildings, particularly in roofs. The use of the relatively inefficient traditional wet insulating materials is therefore likely to decline, in favour of compact, efficient, inherently dry insulants such as plastic foams. However, these materials will only be used successfully if the manufacturers and roof designers fully appreciate the actual environmental conditions which these materials will experience on site and also the misuse to which they will inevitably be subjected. Lack of understanding of these factors has probably made a major contribution to the disappointing performance of plastics materials in roof construction so far, particularly when used for waterproofing. This situation is likely to be exacerbated in the future in conventional roof designs since the roof coverings are likely to be subject to even more thermal stress than at present as the insulation standards are increased.

Roof designers and maufacturers will, in future, need to give much more thought to the concept of the total roof system and not just consider the individual components in isolation.

Therefore, the factors that influence the costs of this element are:
(a) the plan shape
(b) whether or not a pitched roof is suitable
(c) the type of coverings to be used
(d) the life span of the building

Stairs

The cost of the stairs is unlikely, even in the tallest building, to be a major contributor to overall building costs, and their inclusion in a building can hardly be termed an extravagance.

However (and this seems to apply particularly to the finishes and balustrading), it does seem to be an element that falls into that unfortunate category of 'possible savings', and for that reason its cost may well come under close scrutiny. For analysis purposes it should be remembered that the costs can be divided into:

(a) structure
(b) finishings
(c) balustrades

Also when allocating costs it should be borne in mind that for very tall buildings, or even low rise buildings that are to be crowded with people, stairs and stairwells have to provide a high degree of fire resistance.

Although the actual cost of stairs is not usually a critical factor, the positioning of stairs, particularly in high rise blocks, can be. Figure 4 shows two basic alternatives:

(a) with the stairwell/service core integrated into the curtilage of the building
(b) with the stairwell/service core positioned externally.

Alternative (b) will generally be the most economical solution but also the most likely to be rejected aesthetically. (b) is more economical because it produces a better overall gross/net floor ratio, meaning that the overall floor area in (b) can be less than (a) and still fulfil the client's brief, with less area devoted to unprofitable circulation space.

External walls

This is the element in which the architect can impress his personality on the new building, it will determine how the world will view his design and therefore, quite naturally is of great importance to him. It is also an element which contributes a great deal to the overall cost of a building, and one where large areas are involved and an increase of £4 per square metre may sound insignificant until multiplied by 4000 m² of external walls.

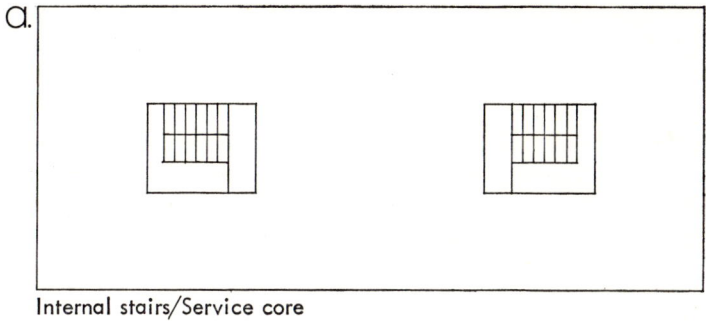

Internal stairs/Service core

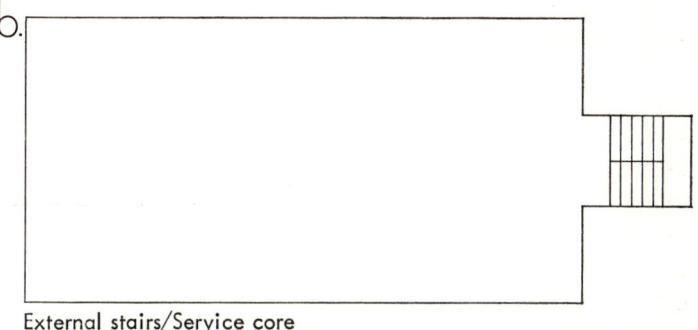

External stairs/Service core

Figure 4

As explained previously the plan shape will determine the wall/floor (or enclosing) ratio, and consequently, the cost of the external walls and windows elements. Figure 6 illustrates how, for an average office block of reinforced concrete frame structure, the construction costs may be expected to be distributed. The plan shape will have an effect on all the largest contributors to cost and especially on the external walls. Therefore, the enclosing ratio is of great importance in the design concept of a building, remembering that the lower the ratio, the better the value for money, and in a well designed office block a wall to floor ratio as low as 0·5:1 to 0·6:1 can be achieved (see figure 5).

For low and medium rise structures load bearing external walls are the most economic solution. Numerous attempts have been made over the last decade to industrialize and prefabricate external wall panels for domestic dwellings. The biggest advantage of prefabrication being, of course, speed. However, traditional brickwork has recently 'hit back' and a pair of traditionally built brickwork semi-detached houses were, according to the builders, completed within eight working days.

For high rise framed buildings, that are to be used as offices or flats, the choice of infill panels and curtain walling systems is vast, the principal governing factor being cost. It is perhaps a symptom of the times in which we live that the availability of building materials

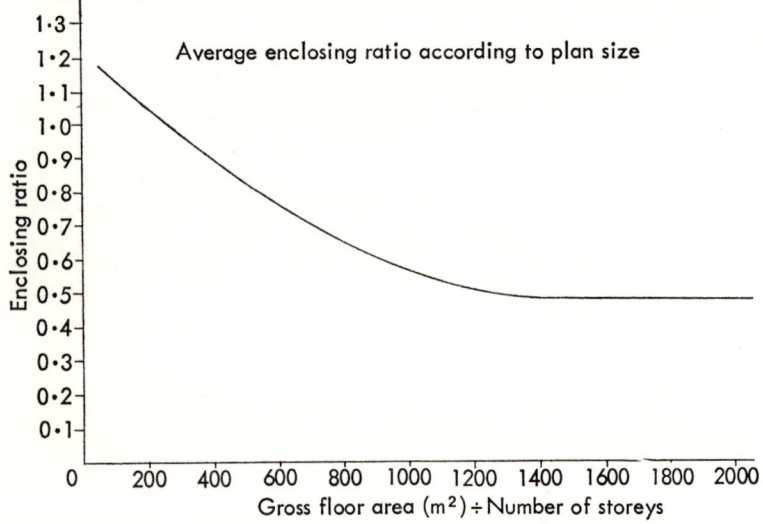

Figure 5

and consequently their cost is cyclical. At the time of writing 'cow coloured' bronze tinted glass is easily available and reasonably inexpensive, with the consequence that a rash of speculative office blocks are appearing incorporating this material.

As the external walls will (irrespective of height or plan shape) envelop the building, their ability to transmit or retain energy is of great importance.

The cavities of cavity wall construction should be filled with insulating foam, and large areas of glass should be double-glazed. These provisions do, of course, add to initial costs (see figure 7 on p. 33), but in the long term will help save and keep down the increasing cost of energy. The decision will have to be taken whether or not to incorporate extra insulation into the structure and the outcome could depend upon the use to which the completed building is to be put. A client, building for owner occupation, will be more concerned about the size of the electricity and heating bills, than if the building is to be let to tenants, when the responsibility will rest with them.

There are so many types of external walling for multi-storey buildings and they have nearly all been introduced so comparatively recently that it has not yet been possible to evaluate the relationship between initial and long term maintenance costs. For example, plate 1 shows a recently constructed office block of a particularly distinctive design.

Preliminaries and contingencies	16%
Substructure	6%
Superstructure	17%
External walls	16%
Services	34%
Finishes and fittings	9%
Drainage and external works	2%

Figure 6

How the façade detail will stand the test of time, what the long term maintenance costs will be, and what it will look like twenty years hence is a matter purely for speculation. Sophisticated systems of curtain walling and fenestration on large structures may contain literally thousands of linear metres of waterproof joints, that will undoubtedly require some form of maintenance, in addition to general cleaning. Therefore, in cases like these, provision should be made for a maintenance cradle at roof level, as the only practical method of carrying out running repairs.

However, using techniques that are fully described in Chapter 5,

Plate 1

a costs-in-use exercise may be reliably carried out on the basis of the materials that have been used for some years now for the external walls of low and medium rise buildings, viz:

(a) Brickwork
(b) Pre-cast concrete
(c) Timber

similar to the exercises included in Chapter 5. Therefore, it may be said that the design criteria to be followed when examining the external wall element are:

(i) Endeavour to produce the most economical enclosing ratio possible
(ii) Use materials that are in 'vogue' and hence easily available
(iii) Use materials with established maintenance costs (but beware! See page 74)
(iv) Incorporate standard details at eaves, external angles, etc.

Windows and External doors

Windows tend to be taken for granted, yet they are an important element of the structure, providing light, ventilation and much of the character of the building.

The design of windows extends to double glazing and double windows which assist considerably towards good thermal insulation and acoustic values.

In the following pages alternative costs of some of the available materials for the manufacture of windows are compared, not only in terms of initial expenditure but also in terms of cost over their life in the building, assuming that a regular pattern of cleaning, repair and redecoration is maintained.

There is a wide choice of materials from which windows can be manufactured, the most commonly used being:

Timber

It is becoming increasingly difficult today to obtain timber of the right quality that has been satisfactorily seasoned. Twisting or warping can lead to badly fitting opening casements with consequent draughts. Failure to protect the timber with adequate weather details can lead to attacks of wet rot requiring subsequent renewal.

Steel
Without adequate maintenance, steel is subject to rust, which gives rise to difficulty in control and to bad fitting.

Aluminium
Aluminium is a metal that does not require protection to prevent it from corroding. With a mill finish, it is subject to chemical action when exposed to the weather and this causes powdering of the surface and pitting of the metal. Frequent cleaning is therefore necessary.

Anodized finish
An anodized finish gives considerable resistance to atmospheric conditions and merely requires regular washing.

Plastic
Solid PVC and fibreglass windows are available, but as yet are comparatively recent innovations. Their qualities and performance therefore have not yet stood the test of time.

Stainless steel
This material is expensive although it should have a long maintenance free life. A stainless steel finish can also be applied to aluminium sections.

Figure 7 gives comparative initial costs for five of the six alternative window materials listed above, stainless steel has been omitted because of its very high initial cost. These have been given in four different groups to reflect changing cost patterns due to size and show how window costs increase when small sizes are used; a fact that does not only apply to the actual window construction itself, but will also be reflected in the higher costs associated with forming the window opening. Comparative costs for double glazing are also given for two of the alternative materials.

Figure 8 compares the total cost or cost-in-use of these alternatives assuming that a regular pattern of maintenance is carried out over a projected 60 year life of the building with future costs having been discounted at 10% per annum compound interest. This 60 year term included in these calculations is liable to considerable change owing to the particular circumstances of each individual.

It will be seen from the final column that the difference in total

Initial Capital costs of various types of windows

Description	Costs per square metre of window area			
	Up to 0·55 m²	0·55–1·10 m²	1·10–1·82 m²	Over 1·82 m²
Standard softwood windows including glass, painting and fixing	£34·68	£33·47	£25·28	£22·64
Ditto but with double glazing	£47·88	£43·26	£33·53	£30·12
Module 100 metal windows including glass, painting, softwood painted sub-frame and fixing	£44·69	£45·57	£30·67	£28·03
Standard aluminium horizontal sliding windows with mill finish including glass, softwood painted sub-frame and fixing	£44·66	£37·51	£25·10	£23·43
Anodized aluminium windows including glass, softwood painted sub-frame and fixing	£42·57	£31·35	£24·53	£21·89
Ditto but with double glazing	£56·21	£43·34	£33·55	£30·80
Plastic framed horizontal sliding windows including glass and fixing (no timber surround)	£102·41	£39·38	£30·91	£25·08

Figure 7

Capital and maintenance costs of various types of windows
Assumed life of 60 years: maintenance costs discounted at 10%

Description	Initial capital cost/m² (1·10–1·82m²)	Maintenance		Total Cost
		Description	Cost/m²	
Standard softwood windows including glass, painting and fixing	£25·28	Wash down and repaint every 5 years. Burn off and repaint every 20 years	£8·67	£33·95
Ditto but with double glazing	£33·53	Ditto	£8·67	£42·20
Module 100 metal windows including glass, painting, softwood painted sub-frame and fixing	£30·67	Touch up and repaint every 5 years. Strip off and repaint every 20 years	£7·39	£38·06
Standard aluminium horizontal sliding windows with mill finish including glass, softwood painted sub-frame and fixing	£25·10	Wash down when cleaning every 6 months. Wash down/burn off and repaint etc. softwood sub-frame	£3·02	£28·12

Figure 8

Capital and maintenance costs of various types of windows
Assumed life of 60 years: maintenance costs discounted at 10%

Description	Initial capital cost/m² (1·10-1·82m²)	Maintenance		Total Cost
		Description	Cost/m²	
Anodized aluminium windows including glass, softwood painted sub-frame and fixing	£24·53	Wash down when cleaning windows every 6 months. Wash down/burn off and repaint etc. softwood sub-frame	£3·02	£27·55
Ditto but with double glazing	£33·55	Ditto	£3·02	£36·57
Plastic framed horizontal sliding windows including glass and fixing	£30·91	Nil	—	£30·91

Figure 8—continued

N.B. If adequate maintenance on softwood windows is not carried out it is likely that they would not last 60 years. Replacement could even occur at 15 year intervals.

cost between the more common alternatives is marginal. If re-decoration is likely to cause inconvenience, disruption or loss of earning potential, the more competitive maintenance reduced alternatives have a clear advantage. If not, where circumstances permit, it might seem prudent to make the selection on aesthetic or other grounds not associated with cost. This element should also come under close scrutiny when considering forms of air-conditioning.

External doors

The selection of material and form of construction for external doors is usually governed by cost considerations, and as with most commodities, the reliability of the product is in direct relation to the initial cost. In a building, this is not a satisfactory state of affairs, as breakdown of basic elements shortly after initial occupation is unacceptable. Where cost limits preclude the use of naturally durable materials for the external fabric of a building, whether doors, windows, or cladding, measures must be taken to ensure that acceptable life expectancy can be guaranteed. With external doors, the most economical assemblies undoubtedly consist of timber constructions of various forms, and it is these which require the most careful consideration at the time of design or selection to obviate failure caused by the inherent weakness of the material. Quality of timber, preservative treatment, protective systems and their subsequent maintenance are all essential to proper functioning for an acceptable period.

Structural failures are less likely where metal or glass components are used, but the architect must satisfy himself that the assembly will meet the necessary standards of performance and that the manufacturers' sales orientated literature does not conceal possible inadequacies in standards of finish and methods of fixing and support.

Internal walls and partitions

The vast choice and cost of internal walls and partition materials range from plastered blockwork to sophisticated demountable partitions, and can be supplied by the client as part of the 'fitting out' contract. Below is an approximate cost guide to various commonly used types of partitions:

Type	Approximate cost m/^{2}
75 mm thick blockwork, plastered and painted both sides	£8·00
Timber stud partition with plasterboard, skim coat and painted both sides	£8·30
57 mm proprietary plasterboard partition and painted both sides	£5·90

Internal doors

As is the case with internal walls and partitions this element does not contribute a large percentage to the total building costs, and listed below are the relative costs of internal door types indicated by a cost index. It has been assumed for the purpose of the index that doors are installed and include wall openings, frame, ironmongery and finishes.

Door type: flush

100 unlipped, hardboard faced, cellular core, softwood frame, painted

105 lipped two edges, sapele plywood faced, cellular core, factory fully finished with clear seal, painted softwood frame

109 lipped two edges, plywood faced, cellular core, softwood frame, painted

118 lipped two edges, plywood faced, semi-solid core, softwood frame, painted

136 lipped two edges, plywood faced, solid core, softwood frame, painted

140 lipped two edges, faced with laminated melamine in standard colours bonded to plywood, cellular core, painted softwood frame, 40 mm thick

200 teak, lipped two edges, hardwood veneered plywood faced, solid core, fully finished, clear seal hardwood frame

127 lipped two edges, plywood faced half hour fire check door, equivalent to BS 459, Part 3, painted

175 lipped two edges, plywood faced, one hour fire check door, equivalent to BS 459, Part 3, painted

Framed

143 40 mm softwood framed, glazed in two panes, painted
210 40 mm hardwood framed, glazed in two panes, polished

Frameless toughened glass

196 10 mm obscure armourcast hung with patch fittings in painted softwood frame, standard latch set

Flexible

403 in one pair, 8 mm black natural rubber panels fixed in painted steel framework with integral double action spring and standard vision panels

Metal

680 polished anodized aluminium box section door and frame, glazed in two panes, overhead concealed spring
807 dull polished stainless steel bonded to aluminium core door and frame, glazed in two panes, overhead concealed spring

Services

An analysis of a typical modern office block will reveal that this major element can, and frequently does, account for up to 36% of total building costs (see figure 6 on p. 25). Because of its increasing importance, somewhat due to the more widespread use of air conditioning, more quantity surveyors and building cost consultants are being asked to measure the services element in their bills of quantities, instead of simply including large and somewhat inaccurate prime cost sums for them. To this end, many large private practices now have their own specialist services quantity surveyors, although bills of quantities are not generally orientated to the measuring of large industrial engineering projects. In fact, unlike building works, there is no convenient, widely accepted standard method of measurement.

For the purpose of this section, and for the ease of this examination only, I have sub-divided services thus:
(a) plumbing and waste
(b) heating and electrical
(c) air conditioning
(d) lifts and special services

(a) Plumbing and waste

A hot and cold water system will be more economic, the more compact it is. For example, in multi-storey structures the kitchens and toilets on each floor should be positioned over each other to avoid long runs of pipework, which will add considerably to costs, and in the case of hot water pipes, only dissipates the heat. There are obvious exceptions where it is not possible to have a compact layout, for example in a school or hospital, and in projects like these the design team should be prepared for costs of services to be considerably in excess of 36% of total costs. Therefore, the costs of the actual pipework, assuming a compact design has been chosen, would seem reasonably static. But the costs of sanitary ware can vary considerably, with specifications ranging from white glazed china with chromium plated taps, to Sicilian pearl marble and gold plated fittings. Here then, is an area where costs should be controlled, and where a possible solution to overspending may be found.

(b) Heating and electrical

The heating installation of a building should be designed in sympathy with the structure into which it is to be incorporated and must be considered on two fronts, initial capital costs and running costs. This is because many of the available systems have widely differing installation costs. From January 1974 to July 1975 electricity charges rose by 76%, and it would seem as though the trend will continue. The cost of fuel oil has also recently risen sharply although by not quite so much, whereas the cost of natural gas has remained fairly constant. Therefore, with possible sharp rises in fuel costs, the correct choice of heating system will be seen to be very important, and a costs-in-use exercise has been included in Chapter 5 to illustrate this point.

When discussing plumbing costs it was said that it was the sanitary fittings that could make or break cost limits. In a similar way, when considering electrical installations the costs of electrical wiring and distribution are reasonably stable, but the cost of fittings, i.e. lighting fittings etc., can contribute up to 30% of total installation costs. In the case of multi-storey buildings where each floor may be let to a different tenant, the architect should ensure that all lighting fittings are installed as part of the main contract and not leave it to each tenant to install his own.

(c) Air conditioning

Air conditioning has, in recent years, become a more common client requirement, whether one is building for owner occupation or for letting. Even so it is mainly restricted to offices/hotels etc. and many installations in the United Kingdom do not compare with the high standards demanded in North America. Until recently special air conditioning was required by all computer installations, but for the next generation of computers the purity, humidity and temperature of the air are not so critical and the general air conditioning plant may be used.

However, even allowing for this somewhat lower standard, if the decision is taken to include this facility in this country, the design team should be aware of the full financial implications, if the system is to function efficiently. It is not simply a matter of allowing for the costs of plant and distribution; there are many other additional associated costs, as shown in figure 10. Also the design philosophy of the entire building should be considered if air conditioning is to be provided, even down to such basic principles as to which elevations will receive direct sunlight and which shade.

The following factors can be said to influence the overall design and cost of air conditioning systems.

(i) *Distribution space*

Assuming that there are five or six major types of air conditioning systems, it should be appreciated when designing that each type requires varying areas for ducting, trunking etc., e.g.

(a) fan coil system will require approximately 7 m² per typical floor of an office block

(b) a dual duct system will require as much as 20 m² per typical floor

In addition it is common practice to conceal most of distribution duct work within the suspended ceiling void, and again each system will require a different height between the suspended ceiling and the slab. In the two systems mentioned above the height required varies from nothing to 450 mm, and this height will of course determine the overall floor to ceiling height, the overall height of the building, the enclosing ratio and overall costs.

(ii) *Plant room space and heights*
The choice of system will be the determining factor in calculating the size and height of plant room required, e.g.
(a) a fan coil system will require a plant room with an approximate total floor area of 325 m² with a clear height of 3·30 m
(b) a dual duct system on the other hand will require up to 530 m² of plant rooms with 4·8 m ceiling heights.

(iii) *Window design, glazing and blinds*
The windows will almost certainly have to be double glazed, and either fitted with blinds, or glazed with sun reflecting glass to maintain temperature levels within the building.

(iv) *Structural module and planning module*

(v) *Building size*

(vi) *Depth of offices (rooms)*

(vii) *Noise insulation*

(viii) *Electrical leads and light fittings*

(ix) *Building location and orientation and resulting shade factors*

(x) *Type of occupancy (i.e. open plan or cellular offices)*

It should be pointed out that these factors vary for every building and cost studies should be undertaken to quantify the effect of them in any specific building project.
When using certain air conditioning systems the positioning of the installation can also be of financial importance, especially in buildings that are to be let to tenants. For example, perimeter heating units should be of sufficient height to permit say, a waste paper basket to be placed beneath them, e.g.

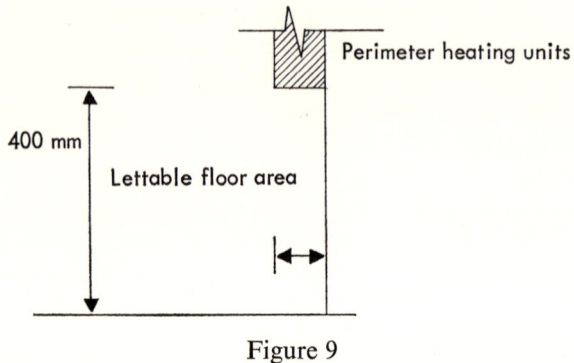

Figure 9

If it is not possible to do this then a prospective tenant may successfully argue that the area beneath the unit, albeit 150 mm wide is unusable and therefore does not warrant including in the net lettable floor area. Therefore this detail, in a large office block with many floors, could substantially effect the gross/net ratio and the revenue.

Figure 10 is an attempt to quantify and cost the additional items associated with air conditioning. The exercise is based on a typical office block of 5000 m², with a basic construction cost of £1 000 000, or £200/m². The type of air conditioning chosen for this exercise is the dual duct system.

The form of air conditioning chosen for the example must be considered to be one of the more expensive systems. Using a cheaper, more simplified installation will produce savings of up to 30% on air conditioning costs, but even this represents a very substantial increase on basic construction costs.

(d) Lift installation

Perhaps the two most commonly used forms of lift installation are electrical traction drive and oildraulic drive. The main advantage for the use of oildraulic lifts is that the ride is smoother and also the levelling at floors is much more precise which is essential for invalid cars, hospitals, etc.

However, the cost of installing the oildraulic lifts together with the necessary bore holes, is approximately double the conventionally driven lift. Maintenance costs are approximately equal for both, but because the oildraulic lifts have no large driving motor, the running

Dual duct system	Cost per m² of floor area (excluding plant rooms) ÷ 5000 m²		TOTAL COST	
	£	£	£	£
Basic office block cost		200·00		1000000
Additional building cost incurred by the introduction of air conditioning				
Air conditioning installation including main contractor's profit and attendance	75·10		375500	
Additional electrical works in connection with air conditioning installation	6·02		30100	
Builder's work in connection with air conditioning	3·88		19400	
Plant rooms at basement level (85 m²)	2·80		14000	
Plant rooms at roof level (445 m²)	14·85		74250	
Duct risers enclosure walls	2·58		12900	
External extract duct housings	2·15		10750	
Venetian blinds on east and west elevations	4·20		21000	
Suspended ceiling within office area	9·26		46300	
Increase in storey height to accommodate air conditioning system above suspended ceiling	6·03		30150	
	£126·87	£200·00	£634350	£1000000

Dual duct system—continued

	Cost per m² of floor area (excluding plant rooms) ÷ 5000 m²		TOTAL COST	
	£	£	£	£
Brought forward	126·87	200·00	634350	1000000
Casings to perimeter heating units	2·15		10750	
	129·02		645100	
Additional preliminaries and insurances for increasing the scope of the works	6·45		32000	
Additional contingency allowance	6·45		32000	
		141·92		709100
TOTAL COST		£341·82		£1709100
Cost per square metre of gross area including plant rooms ÷ 5530 m²		£309·06		
Assuming a gross/net ratio of 76%				
Cost per square metre of net lettable area ÷ 4200 m²		£406·93		

Figure 10

costs during its working life are negligible. A costs-in-use calculation on these lifts has been included in Chapter 5.

As with an air conditioning installation there are many additional associated costs that should be allowed for when preparing cost limits for lifts. These costs range from additional electrical work to work done to motor rooms and general builder's work.

Finishes

This is indeed a vast element and also one which contributes a significant percentage to total costs (see figure 6). Unfortunately, if a solution has to be found for overspending it is usually this element that the design team turn to first. Fortunately in recent years more effective cost planning has given the design team better alternative action. Finishes are generally incorporated in large quantities, and therefore, as with the external wall element, an increase or a decrease of a few pence per square metre affects overall costs considerably. In a book of this type there simply is not the space to discuss in detail the many thousands of different types of finishes. It may seem to the reader to be an easy way out, by saying that each project should be evaluated on its own merits, but nevertheless that is what should be done in this case. An example of such an evaluation has been included in Chapter 5, and some general points on finishes are given below.

For ceiling and wall finishes, dry lining has become very popular in the last decade as it is quick and simple to erect and allows work to progress rapidly. For floor finishes the recent price increase in oil based products means that 'cord type carpets', once thought a luxury, are cheaper to lay than vinyl floor tiles. Computer installations were mentioned briefly when discussing air conditioning, and it should be remembered that such an installation will require a more expensive special raised floor to accommodate the electrical installation.

In America fair faced blockwork has been extensively used in communal areas of flats and other public buildings and because of increasing brickwork and plastering costs, it is becoming increasingly common in this country too. However, to work successfully, the workmanship must be of the highest standard, otherwise the overall effect is catastrophic.

External works and drainage

As with the element substructure, this element is very difficult to evaluate. A multi-storey block on a restricted city centre site will have very little in the way of external works or drainage, whereas a large housing estate in a rural setting may need major sewer and landscaping works. Details of what is required are seldom available until during the contract period and even then the planning authority may alter the layout of the landscaping and the external works. Fortunately this element does not contribute a large proportion towards total costs. However it does contain areas where a choice may have to be made between several alternatives after comparing initial and maintenance costs, for example fencing and road surfacing. Therefore for items like these, it may well be worth carrying out a costs-in-use exercise bearing in mind the disruption that may be caused if large areas of service roads have to be replaced regularly.

Industrialized building

If some distant planet were to observe our traditional building techniques they would think us a very primitive race, constructing the majority of our buildings by cementing one block to another. Industrialized building techniques have been tried and tested for many years, surprisingly without making any significant impact on the construction industry. Perhaps this is even more surprising when, as we have seen throughout this book, traditional methods are labour intensive and labour is a commodity that is rising steeply in price.

In traditional building the sequence of the work and the tasks performed are well understood and the many organizational problems are mitigated by a relatively slow rate of building. Prefabrication in the form of staircases, door sets and trussed rafters, has been successfully assimilated into the process, but some attempts to introduce new techniques or greater prefabrication and a faster rate of building have failed, due to a lack of understanding of the factors which affect productivity on the site. If new techniques are to succeed, there is a need to provide a feedback of information from the site to the effects of innovation, so that design can be continuously modified to improve output.

In Chapter 1, when discussing substructures, the *Finchampstead Project* was mentioned and during this project methods for collecting data of man hour requirements on site were developed and used by the Building Research Station. The technique which has been found to be particularly suitable for detailed studies is activity sampling.

With this technique one observer can study about 100 operatives. Briefly, in an activity sampling study the observer 'snaps' each operative on site at set moments of time, recording the work he is doing at the moment according to a defined code. The following results have been obtained from the activity sampling at the *Finchampstead Project*. The total labour requirements on site expressed in man hours per dwelling were 930, well below the average of about 1 200 man hours for traditional construction.

Average man hour requirements per dwelling

Distribution mains and service connections	60
Site development and external works	310
Substructure	70
Superstructure	90
Services	145
Finishes	255
Total	930

Figure 11

Therefore, if the man hour requirements can be so drastically cut and yet industrialized systems still cannot compete costwise with more traditional methods, then the higher costs must lie with the manufacture of prefabricated components. The capital investment necessary for an industrialized system of building is almost twice that required for traditional methods. This is because many of the prefabricated components have to be made on purpose built factory type production lines and these initial costs are so great that, in the case of housing, it is generally accepted that the minimum number of houses is 50 before the 'break even' point is reached. Before leaving the costs that are associated with labour it is worth noting that although the total number of man hours is lower, a higher proportion of these hours is taken up by skilled erectors and fitters rather than traditional operatives.

Therefore, it would seem that industrialized building techniques can be economically used only under certain circumstances, viz:
(a) low rise housing in schemes of fifty or more units are especially

suitable, provided that there is sufficient time in the pre-contract period to set up the necessary organization,

(b) as so many of the 930 man hours per dwelling in the *Finchampstead Project* were taken up by site development and external works, about one third (310) hours, then the site should be as level and uncomplicated as possible,

(c) the site management staff should be familiar with the system and realize that any problems over constructional details etc., have to be solved immediately and will not have the opportunity to 'come out in the wash'.

It has been said that the use of industrialized building techniques and prefabricated components are bound to increase in the future because of the serious manpower shortages threatening the industry, but, as yet, apart from a few schemes, there seems to be little sign of increased use in these forms of construction in the next decade. However, complacency does not build houses and the Department of the Environment is aware of the present unacceptable time-lag when using traditional design processes, from the initial decision to build to people actually moving into new council homes. Prefabrication will reduce the construction time, as we have seen in the *Finchampstead Project*, but appears to be unacceptable because the associated costs are too high. A large housing scheme will, naturally enough, require a large site, usually bought with borrowed capital at high interest rates and the new houses will not start to produce an income until completion, which could well be years from the date the site was originally purchased. Therefore, with the aim of trying to reduce the total construction period the following procedures are a selection of the ideas that are currently being tried and tested.

Firstly, the Department of the Environment has developed a number of ideas based on the 'develop and construct' procedures. These procedures enable contractors with design skills and constructional techniques to participate in the design process in order to make schemes easier to build and to reduce the design, tendering and construction period while competition between contractors is still maintained. In the Property Service Agency's 'develop and construct' procedure the architect prepares a detailed site layout and selects dwelling types from a range of PSA standard plans based on metric house shells. The selected contractors tender on the basis of the layout and plans, a performance brief and specification clauses, their own method of construction and foundation drawings. The successful contractor then 'develops' i.e. details the design into final working drawings to suit his own method of construction and finally carries out the construction. All drawings produced by the contractor have to receive the architect's approval before use, but the contractor bears the responsibility for design failure.

In the National Building Agency's 'design and build' procedure, the architect prepares a site layout using metric shells and selects contractors with ranges of dwelling plans to fit the metric shells. The contractors tender on the basis of the layout and their own plans, a performance brief and specification clauses.

3 Land Value Determinants

The market price of a building is to a large extent determined by the land values upon which that building is erected. The residual figure, when all the costs associated with the construction of a building, e.g. the actual cost of the bricks and mortar, development costs, consultants' fees, profit, etc., are deducted from the anticipated value of the completed property, is the price that the developer can afford to pay for the land, or the value of the land to him. It is not unusual in certain parts of the United Kingdom for this residual figure to be equal, in monetary terms, to 40% of the building costs. The method of valuation outlined above is known as the 'residual method of valuation' and will be discussed, together with other valuation methods in Chapter 4. The value of land, therefore, and the factors that enhance or detract from its value are of great importance to the design team consultants and building owners.

Before we examine the various methods of determining the value of land it is important to try to define what value is. The actual definition of 'the value of land' seems to be a matter of opinion and has been seen as:

(a) the power to serve man
(b) a relationship of benefits
(c) a basis for choice
(d) an aspect of performance
(e) a measure of efficiency

The term 'value' is usually used in a stock-taking situation or transaction where the worth of an object needs to be expressed in firm, usually monetary terms. However, for the purpose of this book we will consider the value of land as the amount of hard cash that will have to be paid for the freehold or leasehold interest in a plot of land. In whichever way one argues about the definition of land value, there are, undoubtedly, certain factors that influence the formation of any hypothesis.

These influencing factors may be described as:

(a) supply and demand
(b) location
(c) permitted use
(d) a multiplicity of items such as: leasehold or freehold, the availability of needed facilities, such as transport, services, back-up accommodation, etc.

(a) Supply and demand

Many people will respond to the question – what factors influence the value of land? – with the reply: supply and demand. This theory would, of course, readily comply with what appears to be found in practice and could be used to explain, for example, why city centre building plots are highly valued – because the supply is so limited and the demand so great. But, if this were so, then why are not the values in all city centres in the United Kingdom the same? There clearly must be other influences. Yet, despite this last statement I think it must be said that supply and demand does have the greatest effect on building land values, occasionally over-riding all the other factors listed above and sets the general level of values – the other factors may be thought of as 'fine tuning'. For example, during the period 1969–1972 when large amounts of money were available for investment in property, what now, in the light of a more stable market, seem like unbelievably inflated prices were paid for all types of land. Considerations such as those listed above, location, suitability for use, etc., were all swept by the board as people caught up in a 'property owning syndrome' gazumped their way to being 'men of property'. Figure 12 illustrates the staggering increase in the value of land prices from 1969–1973.

No moralizing, but it should be remembered that the people who constitute the supply and perhaps more particularly the demand are not perfect individuals. They are subject to all kinds of pressures, whims and fancies. The most important thing to remember is the wide range and complexity of demand, its unexpected changes, the emergence of new factors and the recession of old, quite apart from

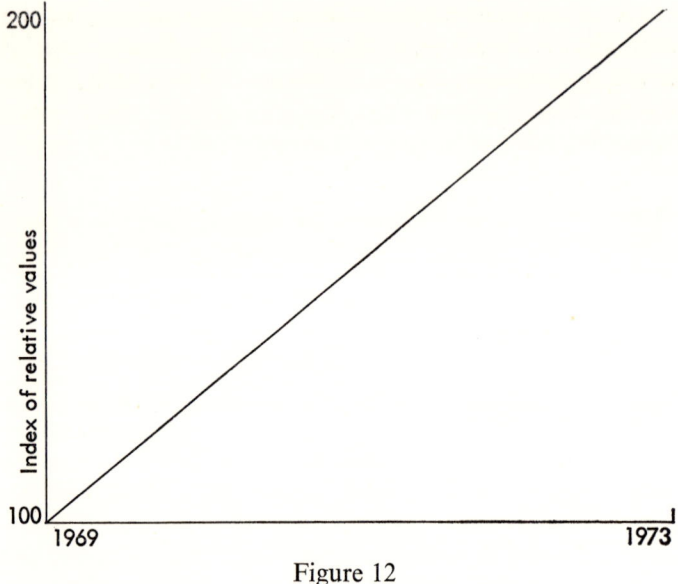

Figure 12

the changes in fashion that make this factor so hard to evaluate and predict. For example demand has been seen to be affected by:

   (i) population shifts, e.g. the recent influx of people associated with the oil industry into Scotland generally and Aberdeen in particular, have caused rapid increases in property values.

 (ii) improved communications

(iii) changes in people's expectations

(b) Location

The influence that the location has on the value of land will depend to what use the land is to be put. For example, if it is to be used for a residential development, then there are a number of elements that people wish to have near to their homes – shops, schools, entertainments, etc., and similarly there are a number of elements that people like to be away from – a large international airport, busy roads, etc. However, to a manufacturer, with world-wide markets, proximity to one of the world's largest airports with all its associated well-developed road and rail links may mean nothing but good business. The relentless roar of the jet engines would probably not influence the value of the land to this type of client. Figure 13 shows an acoustic contour map of London (Gatwick) Airport, showing approximate noise levels. It would be fair to say that generally the higher the noise levels and frequency of aircraft the lower the value for residential purposes.

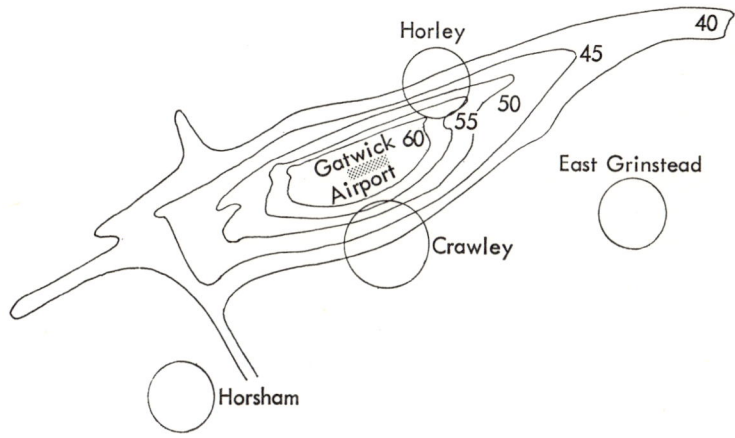

Figure 13

(c) Permitted Use

It is indeed fortunate that, in this country, there is comprehensive legislation that forbids indiscriminate development of our towns and countryside. The aims of the various pieces of legislation are to protect the continuity of our environment and, for example, to ensure that the situation never arises where permission is given to erect an industrial estate in an area of natural beauty or in one that is predominantly residential.

The instruments that control the permitted use of land and therefore influence its value may be listed thus:

(i) planning controls
(ii) listed buildings
(iii) bye-law controls

(i) Planning Controls

The power for local authorities to control the way in which planning is carried on stems from the Town and Country Planning Acts of

1962, 1963, 1968, 1971 and 1972. Generally, planning permission is required for all but the smallest structures and applications may be in outline, giving only brief details of the expected use in order that permission may be secured in principle, or they may be detailed. A detailed application is required for every development affected by the act. The reason why outline planning permission is sometimes sought is so that a plot of building land may have its value increased once permission has been granted, and therefore become a more easily saleable commodity; also the documentation necessary for an application for detailed planning permission is considerable. Once granted planning permission is valid for five years and construction must be commenced within this period, or another application is necessary. Planning proposals for city or town developments will not only show the permitted use of areas contained within the actual project but enable an overall planning concept to be possible by indicating the permitted uses of adjoining areas. This power must be exercised with great care as the re-planning of, say, a city centre can quite easily alter the pattern of land values. Not only will a map show the permitted uses of land but also will state the densities to which they may be developed. These densities will be referred to again in connection with the *Housing Cost Yardstick* calculations in Chapter 4. In addition to the planning acts there is a complementary act known as the 1973 Land Compensation Act, the aim of which is to compensate for what is termed 'planning blight'. It is possible, as discussed earlier, to change property values by planning decisions, for example, if a new motorway were to pass by a plot of building land, thereby causing a potential nuisance and a drop in its value it is possible for the land owner to be compensated for this. If the motorway were to pass through his land then it is likely that it would be bought from him compulsorily.

(ii) Listed buildings

The Department of the Environment has compiled a list of historic buildings and buildings of interest in an attempt to preserve them. These buildings cannot be demolished or altered without prior application to the department. Any alterations must be in keeping with the original structure. This list, of course, is not a complete answer, as it has been known for listed buildings to be demolished either in ignorance or by people who chose to turn a blind eye to the list's existence.

(iii) **Building regulations**

As well as administering the Town and Country Planning Acts the local authorities have responsibility for ensuring that the method of construction to be used in a proposed building comply with the 1972 Building Regulations. The regulations aim to lay down minimum standards to ensure such things as structural stability, fire resistance and materials.

(d) Freehold or leasehold

Strictly speaking, in the eyes of the law, it is impossible to purchase outright a piece of land. One is only able to buy a freehold interest in that land, as all land in this country is held by the Crown. Nevertheless, ownership of the freehold of a plot of land does mean that a person is entitled to do as he wishes on the land, so long as he does not break the law or prove a nuisance to adjoining owners.

One of the options open to a person who holds the freehold of a piece of land is to lease part, or all, of it to tenants in return for remuneration of some kind. If it is decided to lease part of the land then the terms of the lease can have an influence on the value of the land, as they generally lay down the way in which the land may be used. The length of the lease is important; the shorter the term, the lower the market value. Most agreements are for periods of 99 years. The lease may also contain restrictive covenants and these should be studied very thoroughly as they lay down the ways in which the land may be used by the lessee. For example, a building plot that is large enough to have two blocks of flats constructed on it may have a restrictive covenant in the lease permitting the land to be used only for the construction of a dwelling house. Clearly the return that a builder could obtain from a single dwelling house as compared with two twelve storey blocks of flats would vary considerably and therefore the value of the land would be affected accordingly.

I have also grouped together in this final category such common sense items as the availability of transport facilities; isolated plots of land usually have lower values than those with well developed transport links, no matter what the intended use may be. Another factor that may also be included within this group is the topography of the land, because quite clearly the uses to which a steeply sloping site may be put are limited. On top of the physical problems of getting the best out of the land from the designer's viewpoint there is usually the added cost of construction, as almost certainly stepped foundations of even steel sheet piles as permanent retaining walls may have to be used.

Perhaps now that the reader has given thought to the items which influence the pattern of land values he is in a better position to formulate for himself a definition of the value of land.

4 Methods of Valuation/Budget Planning for Local Authority Housing

Methods of Valuation

Now that we have endeavoured to define the term 'value' and to equate the factors that influence it, the next step is to examine the methods of valuation available to fix a price on a piece of land, or property. The most commonly used methods are:
(a) residual method
(b) investment method
(c) comparative method
(d) profits method
(e) contractor's method

As with any technical process, there are certain basic principles that, in this case, the valuer should be familiar with, before he is able confidently to try to establish the value of a piece of property. The terms and formulae required in the valuation process are explained in the following paragraphs; some of them will have been briefly explained in Chapter 2. The information that is included in this chapter will enable the building economist to calculate, say, the *present value* of £1, or the *amount* of £1 should the need arise. However, in practice it is not necessary to perform such lengthy calculations for each problem, as reference to a set of valuation tables, such as *Parry's Valuation Tables*, that contain all the data required in tabulated form, is generally the easiest and most reliable source of information. See appendix B at the back of this book for an example of valuation tables.

The present value of £1 per annum, or years purchase

If a freehold interest in land were bought for £10000 by a client who requires a 15% return on his investment, then this can only be achieved if the interest that he receives yearly is:

$$£10\,000 \times \frac{15}{100} = £1500$$

(capital value)   (net income)

By using this information in reverse it is possible to calculate the capital value when the other pieces of information are known.

$$£1500 \times \frac{100}{15} = £10\,000$$

The expression in this example, 100/15, is known as the *years purchase*, or the *amount* of £1 per annum. Therefore, it can be said that *net income × years purchase* = capital value, and clearly the *years purchase* can be calculated by dividing 100 by the *amount* of return on capital required, which in this case would give 6·67. The amount of return on investment will alter from one client to another and will depend upon the amount of risk involved and the interest rates that are available if the capital were invested, at less risk, elsewhere.

Example
Value a freehold interest in a property producing a net income of £5000 p.a. The client requires a return of 12% on his investment.

$$£5000 \times \frac{100}{12} = £41\,667$$

The above example concerns investments in perpetuity, i.e. a freehold interest in land, the land will be in the possession of the purchaser for perpetuity unless, of course, resold, in which case the capital is recouped. For acquisitions or investments with a limited life, e.g. a leasehold interest in land, it is necessary to invest each year a sum, paid out of income, into a sinking fund. These annual sinking fund instalments accumulate at compound interest over the term of the lease to a sum that equals in amount the original capital sum invested. Thus, by the time the original capital sum has wasted away, a fresh sum of the same amount is ready to replace it. When dealing with the valuation of wasting assets, dual rate valuation tables should be used, as they contain a built-in sinking fund allowance.

It should also be remembered that with inflation currently running at over 25% per annum, it will not be possible to purchase real estate at today's prices in 20 years, and an allowance should be made for this, or at the very least the client should be informed that all figures quoted are based on today's costs, with no allowances for inflation.

Annual equivalent in perpetuity

On a 10% basis, the interest that would be earned in a 12 month period on the sum of £100 is £10. Therefore, the annual equivalent of £1 invested at 10% in perpetuity is £10/100=£0·10.

The idea of annual equivalent is akin to the theory of opportunity costs, as it represents the amount of interest that will be forgone if, for example, a sum of money is invested in property, instead of being invested at a given rate per cent of interest.

For example, calculate the annual equivalent on a 10% basis of £150000 which is the price paid for the freehold interest in a plot of land.

purchase price	£150000
annual equivalent of £1 in perpetuity at 10%	0·10
	———
Annual equivalent	£15000
	———

The annual equivalent of a capital sum can be calculated using the following formula:

$$i = \frac{R\%}{100}$$

where, i = annual equivalent

R = rate per cent of interest

Annual equivalent of wasting assets

When considering *years purchase*, it was explained that when investments have a limited life it is prudent to make allowance for a sinking fund. Therefore, if instead of investing the sum of £150000 mentioned earlier in a freehold interest in land it had been invested in a property with an expected life of 40 years, then, in addition to calculating the annual equivalent as before, a further allowance will have to be made to replace the capital in 40 years' time.

The annual sinking fund to replace £1 in 40 years at 10% can be calculated thus:

$$\frac{i}{A-1} \quad \text{where} \quad i = \frac{R\%}{100}$$
$$A = (1+i)^n$$
$$n = \text{number of years}$$

$$\frac{0\cdot10}{(1+0\cdot1)^{40}-1} = 0\cdot0023$$

For accuracy, the annual sinking fund instalment must be adjusted to take account of the effects of income tax, which otherwise would prevent the annual sinking fund from accumulating to the full amount required. This can be done as follows:

$$\frac{\times 100}{100-P} \quad \text{where} \quad P = \text{the client's rate of income tax}$$

Therefore to calculate the annual equivalent of £150 000 which has been invested in an asset with an expected life of 40 years, the annual equivalent calculated previously, 0·10, must be added to the annual sinking fund allowance of 0·0023 to give a total of 0·1023. To avoid clouding the issue the effect of income tax on the annual sinking fund has been ignored.

$$\begin{aligned} &£150000 \\ &\times 0\cdot1023 \\ \hline &\text{annual equivalent } £15345 \\ \hline \end{aligned}$$

The amount of £1

Using this formula it is possible to calculate the *amount* of compound interest that will accrue to a single initial deposit of £1 over a number of years. No subsequent deposits are made. The formula gives the total of the interest and the initial deposit.

$$A = (£1 + i)^n \quad \text{where } i = R\% \text{ as a decimal of £1}$$
$$n = \text{number of years over which the sum is to be invested}$$
$$A = \text{the amount of £1}$$

The present value of £1 formula

This formula can be used to calculate the sum that must be invested now to accumulate to £1 at the end of a stated period, at a given rate per cent. The formula is:

$$V = \text{the amount needed to come to £1}$$

$$V = \frac{1}{(1+i)^n} \quad \text{where } i = R\%, \text{ as before}$$

$$n = \text{number of years}$$

These are the 'tools of the valuer's trade'. The following examples of the various methods of valuation will show how they are used and their inter-relationship.

(a) The residual method of valuation

This method of valuation involves calculating the *gross development value* of a building scheme, or the market price that it is expected to realize when the land has been developed and disposed of, by selling or leasing, and then deducting from this GDV all the costs that will be incurred during its development, including the developer's profit. The residual figure represents the amount that it is possible to pay for the land, in order that it can be developed and disposed of at a profit.

Example
How much could your client afford to bid for 3 hectares of land with planning permission for detached houses at 25 per hectare? Houses of a similar type realize £26 500 in the vicinity.

	£	£
Gross development value		
3 hectares @25 houses/hectare =		
75 houses selling for £26 500		1 987 500
Less		
Costs of construction		
75 houses @ 95m² each @£160/m² =	1 140 000	
roads and sewers (say)	36 000	
	1 176 000	
Architects', Q.S. and Consultants' fees 10%	117 500	
	1 293 500	

	£	£
Interest on borrowed capital		
say £646 750 for 2 years @ 14%	181 090	
Legal, agents' and advertising		
fees (for sales) @ 3% of GDV	59 625	
Developer's profit 15% of GDV	298 125	
	———	
TOTAL COST	1 832 340	1 832 340
		———

	£
Sum available for purchase of site × PV £1	
(0·7694675) in 2 years @ 14%	155 160
(This approximately represents the interest	
on the money borrowed to buy the	
site plus lawyer's fees incurred in	
purchasing it)	
£155 160 × 0·7694675 = £119 390 say	£120 000
	———

which is equal to £40 000 per hectare or £1600 per plot, which represents the amount that the client can afford to pay for the land in order to develop it and make a profit. It should be noted that construction costs and interest on borrowing depend on the client. The residual method of valuation is not so much a valuation, as an estimate of how much a client could afford to bid.

(b) Investment method
This method is used where the property produces an income, e.g. a shop. The income from the investment in the shop must prove to be more profitable than investment in, say, a building society.

Example
Value the freehold interest in a shop producing a net income of £3000 p.a. The purchaser requires an 8% return on his capital. Remember from previously that

$$\text{capital value} = \text{net income} \times \textit{years purchase}$$

$$\text{therefore capital value} = £3000 \times \frac{100}{8} = 12\tfrac{1}{2}$$

$$= £37\,500$$

This example is based on investments in perpetuity; leasehold investments require an allowance for a sinking fund.

(c) Comparative method

Perhaps the most simple of all methods, this process involves a direct comparison with similar types of properties to the one being valued in the vicinity. The price paid on the open market for comparable properties forms the basis for fixing a price.

Residential properties are, in the main, valued on this basis with additions or omissions being made from the price paid for the comparable property for such things as:

(a) rear extensions
(b) standard of decoration and fittings
(c) aspect
(d) central heating (if any), etc.

Local knowledge is also useful. For example, in Hastings, Sussex, on what is perhaps unkindly referred to as the 'Costa Geriatrica', because such a high percentage of the population and the house buying public are over retirement age, a house with a long steep flight of steps to the front door would probably be valued lower than a house having easy access.

On a housing estate, comprising many hundreds of houses, with perhaps only five different house types, the process of comparative valuation is a relatively simple affair and indeed householders can quite easily, with a little local knowledge, value their own property. However, in more distinctive areas with many individual properties, the valuer would probably have to split up, or zone the property into meaningful units of comparison.

(d) Profits method

This method is used for properties that have an earning capacity, for example, theatres, clubs, etc. It involves establishing the gross earnings of the property and deducting from this all expenses, including profit, that are likely to be incurred by the tenant. The residual figure is the amount available for rent.

(e) Contractor's method

This method is based on the theory that the value of a building and the land on which it stands is equal to the cost of construction plus the value of the site. This statement is clearly not true, as already in a previous chapter we have seen that the value of a building is the price that people are prepared to pay for it on the open market. Therefore, the idea that this method of valuation is one favoured by contractors is not true, the only time it is used is for properties that rarely are offered for sale on the open market, such as schools, hospitals, etc.

Budget Planning for Local Authority Housing Schemes

Housing Cost Yardsticks

In 1967 the Department of the Environment issued a circular to local authorities entitled *Housing Standards, Costs and Subsidies*. This document, together with all its numerous subsequent revisions and amendments, forms the basis of the *Housing Cost Yardstick*, which lays down the procedure for obtaining loan sanction and Treasury subsidies for local authority housing, as well as establishing the standards and the cost limits with which housing must comply. The *yardstick*, a copy of which may be obtained from any branch of H.M. Stationery Office, contains the necessary data for calculating the basic cost limit for proposed housing schemes by means of a series of tables setting out the cost limits for a wide range of situations. The following text includes an example of how a *yardstick* submission could be presented, but it should be remembered that although all submissions for approval are made on a special form (HCY1) there are no hard and fast rules and it is sometimes a matter of judgement as to how some of the calculations are presented. Also the *yardstick* tables on which the calculations are based are those current at the time of writing.

Before the start of the example it would be helpful if the reader could be armed with a copy of the latest *housing cost yardstick* while working through the chapter. Firstly, for the purpose of using the *yardstick*, the cost of a building project is divided as follows:

(a) Items which form part of the dwelling itself, i.e. the substructure, the superstructure and certain external works

(b) Items that relate to the building but do not qualify for subsidy, e.g. TV aerials, etc., the total cost of the groups should not exceed 10% of the agreed *yardstick*

(c) The third category of costs are those that relate to garages, car-parking, site development works, etc. Except in the case of roads and sewers, these items do not qualify for subsidy and there is a separate table setting out the cost limit for car accommodation.

In order that the *yardstick* tables may be applied to all housing schemes there have been built into the tables three variables:
(i) the average number of persons per dwelling
(ii) the density of the development. This is necessary because different local authorities have different ideas about the ideal planning density for housing, and of course virtually every building site will be capable of being developed to a different density.
(iii) allowance for the regional variation in building prices throughout the country.

(i) Average number of persons per dwelling

The size of a dwelling is expressed as the number of persons, referred to as bed-spaces, per dwelling. The *yardstick* requires that all calculations are based on the average number for the whole development, i.e. the total number of bed-spaces divided by the total number of dwellings, and not on individual dwellings.

(ii) Density

The density of a scheme is expressed as the number of bed-spaces required per hectare of the site. The site area is for *yardstick* purposes, the land to be used for building with the boundary taken at half the width of any roads bordering the site.

(iii) Regional variations

Because of the vast range in the levels of building costs throughout the country, it is necessary to adjust the level of the subsidy that is available to allow for this. The *yardstick* contains a list of additions to the basic costs, expressed as a percentage, and these should be used in all calculations. Figure 14 shows the regions of cost variance and the percentage additions that are applicable to them.

Example

The following then is an example of how the preparation of a *yardstick* submission, for a hypothetical scheme, could be approached, but first here is a list of basic information that the building economist will require before he can start work.

Area of site (for *yardstick* purposes)   1·04 hectares
Number of dwellings   105
Total number of people (bed-spaces)   344
Car spaces   64

REGIONAL VARIATIONS
To the yardstick costs

Map showing the
percentage adjustments
applicable in the various
Economic Planning Regions,
for variation of price
contracts.

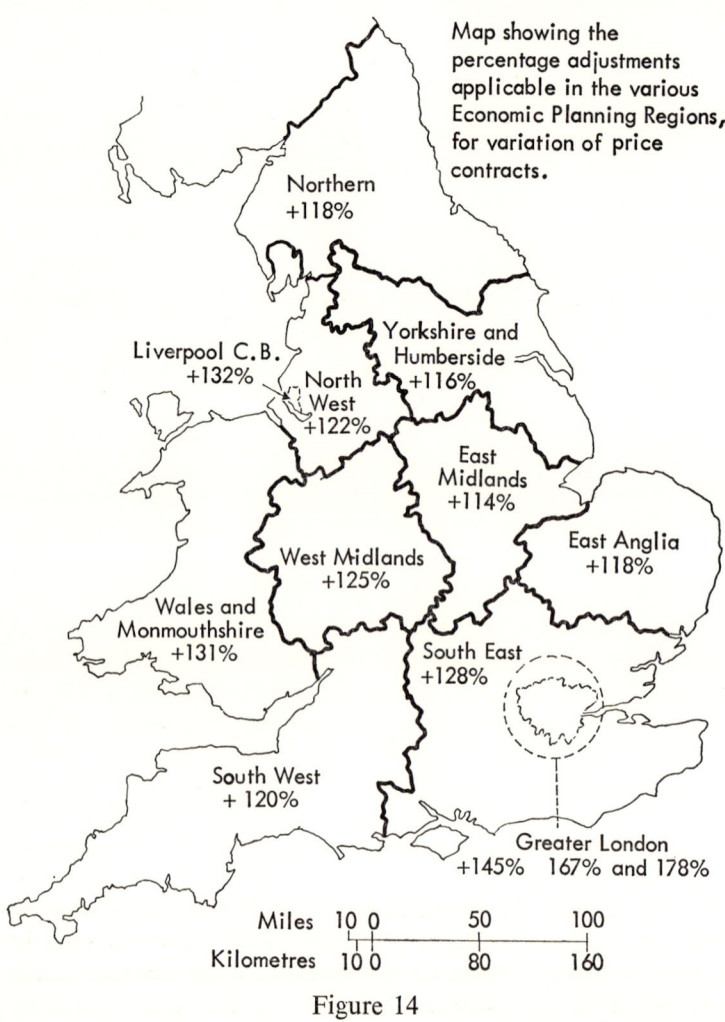

Northern
+118%

Liverpool C.B.
+132%

Yorkshire and
Humberside
+116%

North
West
+122%

East
Midlands
+114%

West Midlands
+125%

East Anglia
+118%

Wales and
Monmouthshire
+131%

South East
+128%

South West
+ 120%

Greater London
+145%   167% and 178%

Miles   10 0   50   100
Kilometres 10 0   80   160

Figure 14

From this information the density of the development can be found as follows:

$$\frac{344}{1 \cdot 04} = 331 \text{ persons (bed-spaces) per hectare}$$

and the average number of bed-spaces per dwelling as $\dfrac{344}{105} = 3 \cdot 3$ this

information will be required again later when consulting the *yardstick* tables.

The building economist should now, using his own cost information, estimate the cost of the proposed scheme, as a check against the amount of the level of subsidy it will be able to obtain.

		£
1	Site clearance, breaking up hardstandings, etc.	60 000
2	Dwellings	
	(a) Houses 1170 m² @ £100	117 000
	(b) maisonettes 1015 m² @ £100	101 500
	(c) maisonettes and flats with garages 3680 m² @ £130	478 400
	(d) old persons' flats 4310 m² @ £120	517 200
3	Anti-flood measures	100 000
4	External works and drainage	130 000
5	Noise prevention	60 000
6	Design risk	30 000
		£1 594 100
	Say	£1 600 000

As the above calculation shows the development for the purpose of this example has been taken as a typical housing scheme with a mixture of houses, maisonettes, flats, together with some accommodation for old people, the exact requirements for permitted planning densities should be established with the local authority beforehand. Items 3 and 5 have been included as an example of typical *ad hoc* items that may be encountered. Now that the approximate cost of the scheme has been established it can be seen how it compares with the *yardstick* allowances. Based on the information that has been calculated earlier in this example the *yardstick* allowance, found by reference to the tables, for a development with 3·3 bed-spaces per dwelling to a density of 331 persons per hectare is £971·00.

This allowance of £971·00 will be shown in the tables as three separate sums:

	£	£
(a) superstructure	848	
(b) substructure	60	
(c) external works	63	
	£971	
£971 × 344 (number of people)		334024

In addition to the houses, flats and maisonettes in this scheme, there are to be 53 dwellings especially for old people. The *yardstick* recognizes that the building of such special dwellings involves extra expense and accordingly gives an added allowance, as follows:

26 one person dwellings @ £415	10790
27 two person dwellings @ £480	12960
	£357774

This figure now represents what the *yardstick* calls the normal level of sanction. However there are further allowances to add in, first, as mentioned previously, an allowance for regional differences in pricing. In the case in question the area in which the project is to take place qualifies for an addition of 50%

£357774 × 50% = 178887	178887
	£536661

The next consideration is what is termed 'exceptional costs' which qualify for *ad hoc* additions. These costs should already include the regional variation allowance. The 'exceptional costs' are allowable under the following headings:

    (a) special design relative to setting
    (b) use of special materials
    (c) special foundations and/or external works

Under this heading an addition of £195 000 has been included for which substantiating calculations will have to be submitted. These are contained in Appendix 1 to this example on page 67

	£
	195 000
	£731 661

This total is termed the 'exceptional limit'; however, there are still other allowances that may be claimed. Car accommodation is allowed for by making reference to the *yardstick* tables once more, as follows:

32 car spaces @ £50	1 600	
32 garages @ £375	12 000	
	13 600	
Add 50% regional variation	6 800	
	20 400	20 400

Finally, the *yardstick* allows a tolerance of 10% on, in this case, the exceptional limit total which is included for loan sanction, but not for subsidy

10% × 731 661	73 166
	£825 227

The *yardstick* form also requires, for information only, any other costs that are to be incurred in the building works but for which *yardstick* allowances are not given, e.g. site development costs, say £140 000. As before substantiating information is submissible (see Appendix 2 page 67).

Now that both the estimated cost and the *yardstick* allowances are known the two may be compared thus:

Estimated Cost		Yardstick Allowances
		£
£1 600 000	dwellings	731 661
	tolerance costs	73 166
	car accommodation	20 400
	site development	140 000
		£965 227

Therefore, the amount by which the building economist's own estimate of cost exceeds the *yardstick* allowance is almost £650 000. The gap is clearly unacceptable and could be even greater if the *ad hoc* additions are not approved. The main reason for the large difference between the two estimates is the fact that building costs have risen so rapidly since the publication of the *yardstick tables*. So steep was the increase in building costs that for two years (November 1972 to October 1974) the Department of the Environment was prepared to treat each application on its own merit, as it recognized the 'complex and erratic tendering situation which had developed'.

Then in October 1974 new regional variation to cost tables were published which showed the amazing *yardstick* plus percentages on figure 14. Therefore, assuming that the building scheme being considered is in Croydon, the *yardstick* percentage will increase from 50% to 167%. Repeating the calculation on this basis gives the following answers:

Estimated Cost	Yardstick Allowance
£1 600 000	£1 550 000

which are clearly more favourable

After the project has been out to tender and the lowest one chosen yet another *yardstick* form, known as a TC2, is completed, on the basis of the information in the lowest tender, and submitted to the Department of the Environment for comparison with the HCY1. If it becomes apparent that there are any unexpectedly large or additional costs, then the Department may listen with a sympathetic ear and grant a higher level of subsidy.

Appendix 1
Exceptional 'ad hoc' *additions*

	£
1 Extra cost of anti-flood measures	130000
2 Extra cost of special foundations	20000
3 Extra cost of noise prevention	40000
4 Extra cost of accommodation for disabled old persons	5000
	£195000

Appendix 2

	£
1 Site clearance, breaking up existing concrete	60000
2 Roads, nameplates, drainage and adjacent paths	20000
3 Communal footpaths	5000
4 Grass and landscaping to public areas	14000
5 Street lighting	3000
6 Work to existing services	38000
	£140000

5 Design Evaluation Techniques

Throughout the preceding chapters we have said that when deciding on the types of materials to use in a building it is not merely a matter of comparing 'face value' costs. For example, figure 7 illustrates the overall comparative costs associated with various sorts of windows. To obtain a true picture of the various design alternatives not only do the initial costs of a form of construction or type of installation have to be taken into account, but also the subsequent running and maintenance costs. This overall comparison of costs can be performed by considering all the costs that are associated with building; for example:

(a) the cost of the land (if it is freehold, then it will be a capital sum or, in the case of a leasehold plot, an annual rent – an annual sum)

(b) the costs of construction together with all the associated professional fees (a capital sum)

(c) annual running costs, for example heating and annual maintenance (an annual sum)

(d) periodic expenditure for regular maintenance, for example boiler replacement, replacement of roof covering (capital sums to be met at regular intervals)

(e) a further cost may be a premium paid to a landlord every tenth year. Such a premium means that a lower rent is asked and therefore the annual equivalent is already taken into account as a deduction from the full market rent.

Using techniques that will be described shortly, all the capital payments listed above are reduced to their annual equivalent and then to this figure are added annual running costs, outgoings or maintenance costs to find the total annual equivalent of the various proposals. This method of comparison is useful to the client as it is possible in this way to compare annual liabilities with expected income.

Although the comparison of design alternatives using costs-in-use techniques is of great value, it would be naive to think that it is the ultimate answer to cost forecasting. The disadvantages can be summarized as follows:

(i) There are difficulties in obtaining information on the maintenance and running costs of various materials and systems. The Building Maintenance Cost Information Service of the Royal Institution of Chartered Surveyors has been in operation now for several years, but is still in its infancy and needs years more yet before it can hope to show any long term trends or economies in using particular materials. The service is similar in character and organization to the Building Cost Information Service in so far that maintenance costs are analysed along with a description of the age, construction and maintenance organization for the analysed building. Data sheets of analyses of the Property Occupany Costs of individual buildings are prepared from information submitted by BMCIS subscribers as part of the reciprocal nature of membership. Occupancy Cost Analyses are designed to fit in with an organization's internal procedures and the analyses build up a library of maintenance cost data which will help property and maintenance managers to compare their own operating circumstances and occupancy costs with those of other organizations. Appendix C (see page 89) shows a typical BMCIS Occupancy Cost Analysis. The analysis also breaks the costs down into elements, viz:

Improvements and Adaptations

Work arising from building extensions, alterations and changes in use or quality of buildings (but not direct replacements which would be included with the appropriate element below). Included here would be new construction work, alterations to structures or services, installation of new services and work to comply with legal requirements.

e.g. new annexe, moving of partitions, alterations to heating services, construction of fire escape, etc.

1 Decoration

1·1 External decoration: decoration or redecoration externally. Include any making good and washing down of any decorated surfaces in lieu of redecoration such as washing down walls. Exclude washing down undecorated external walls, such as stone-faced elevations which will be included in 'element 4·2 cleaning external surfaces'.

1·2 Internal decoration: decoration or redecoration internally. Include any making good and washing down of any decorated surfaces in lieu of redecoration such as washing down walls.

2 Fabric

2·1 External walls: repairs to external structural walls, curtain walls cladding, glazed screens, external doors and windows; restoration (e.g. stone work) should be included here.

2·2 Roofs: repairs to flat and pitched roofs, roof lights, lean-to roofs and the like, together with associated work such as roof flashings, dpc's, gutters and downpipes. Where costs of work to pitched and flat roofs have been recorded separately these can be shown separately.

2·3 Other structural items: repairs to ducts, internal doors, borrowed lights, frames, stairs, balustrades, dado rails, gantry rails, ironmongery (e.g. locks and closers), external fire escapes, floor structures, etc. Where costs of work to any of these items have been recorded separately, they can be shown separately.

Exclude heating ducting which is under 'element 3·2 heating and ventilating'.

2·4 Fittings and fixtures: repairs to fitted cupboards, seats, notice boards, shelving, worktops, fireplaces, grillages, blackboards, etc. Loose furniture such as chairs, desks, etc., furnishings such as curtains, drapes and blinds, etc., and any fittings associated with mechanical or electrical services such as cookers, and free standing fires are not to be included.

2·5 Internal finishes: repairs to internal finishes such as floor finishes (tiles, blocks, carpeting, etc.) wall finishes (plaster, lining board, tiles, etc.) and ceiling finishes (plaster, suspended systems, tiles, etc.). Exclude making good any surfaces for decoration.

3 Services

3·1 Plumbing and internal drainage: repairs and servicing to plumbing and internal drainage including work to rising mains, storage tanks, and cisterns, hot and cold water services, sanitary ware, WC pans, urinals, sinks, taps, valves, waste, soil overflow and vent pipes, internal manholes, rodding eyes, and access covers, and the cleansing of interceptors.

Include work to hot water pipework and accessories from the calorifier or hot water tank to outlets. Exclude calorifiers, expansion tanks and the remainder of the primary system which are included in 'element 3·2 heating and ventilating'. External drainage is included in 'element 8·2 external services'.

3·2 Heating and ventilating: repairs and servicing to fuel tanks, boilers, flues, plant, pumps, motors, filters, switches, expansion tanks, pipework up to and including calorifiers, radiators, ducts, valves, fans and equipment associated with heating and ventilating installations and air conditioning.

3·3 Lifts and escalators: repairs and servicing to lifts and escalators including paternosters and dumb waiters.

3·4 Electric power and lighting: repairs and servicing to electrical switch gear, fuse boxes, busbars, casings, wiring and conduit to lighting and power supply.

Exclude special circuits (starting from their controlling fuses and switchgear) such as fire alarms, bell systems, emergency lighting, clock systems, etc.

3·5 Other mechanical and electrical services: repairs and servicing to other mechanical and electrical services which are part of the building, such as filter plant, fire alarm and bell systems, emergency lighting, clock systems, hospital gas (oxygen, etc.) lines, fire fighting equipment, incineration, compressed air equipment, cold stores, and special electrical services, flood-lighting, lightning conductors.

Exclude user services which are independent of the building such as telephones, typewriters, cookers and refrigerators, and cranes, unless forming part of the function of the building, e.g. overhead gear in abattoirs.

4 Cleaning

4·1 Windows: cleaning windows, glazed curtain walling and glazed screens etc. State in the 'brief description' the frequencies and whether one or both sides are cleaned.

4·2 External surfaces: cleaning external surfaces of the building, e.g. cladding or stone work face, etc.

Exclude the cleaning of external walls surfaces which is accompanied by restoration (e.g. stone work), which is included in 'element 2·1 external walls'.

4·3 Internal: all internal cleaning, other than windows, e.g. cleaning floors, vacuum cleaning, shampooing carpets, dusting and cleaning ledges, furniture and fittings. Include all costs of materials and allocate machine costs.

Exclude washing down walls or paintwork done in lieu of redecoration, which is included in 'element 1·2 internal decoration'.

5 Utilities
The net costs arising from the use of utilities should be recorded under one of the sub-elements.

Exclude attendants' costs which are included in 'element 6·1 services attendants'. Where costs are applicable to several buildings as when several buildings are served by only one meter, or there is a district scheme, costs should be allocated by floor area ratios and the detail shown on an accompanying sheet. Sub-elements are as follows:

5·1 Gas

5·2 Electricity

5·3 Fuel oil

5·4 Solid fuel

5·5 Water rates

5·6 Effluents and drainage charges

6 Administrative costs
6·1 Services and attendants: salaries, wages, insurances, etc., and other costs of boilermen and such labour attending upon the equipment directly serving the building and including odd-job men whose costs cannot be allocated to specific functions and also caretakers.

Porters (element 6·3), security staff (element 6·4), staff executing rubbish disposal (element 6·5) and management staff (element 6·6) are not included in element 6·1.

6·2 Laundry: provision of toilet facilities (e.g. towels, soap, etc.) relevant to the functioning of the building and including cleaning uniforms of services attendants.

Exclude special user requirements such as the laundering of hotel bedding, etc.

6·3 Porterage: salaries, wages, insurances, etc., and other costs of staff engaged fetching, carrying and directing visitors.

6·4 Security: salaries, wages, insurances, etc., and other costs of staff engaged to maintain the security of the building, or payments made to a contractor executing this service.

6·5 Rubbish disposal: salaries, wages, insurances, etc., of staff engaged in the collection, removal and disposal of rubbish from buildings, together with an allocation of cost for any equipment used, or payments made to a contractor executing this service.

6·6 Property management: all costs of supervisory staff (e.g. architect, surveyor, building supervisor, etc.) and clerical staff.

Note:
Where it is appropriate to make an allocation of cost, this should be as set out in the 'Principles of Analysis'. Administrative costs include expenses such as travel, car allowance, overtime payments, etc. but care should be taken not to include the cost of DEL forement, etc. whose costs are considered direct overheads to the DELF (see notes referring to 'Percentage additions to cover overheads'). Costs should also include an allowance for the use of equipment where this is necessary and these should be allocated as set out in the 'Principles of Analysis'.

7 Overheads

7·1 Property insurance: premiums for insuring the property. A description of the insurance cover, all properties covered and the capital sums insured, should be stated.

7·2 Rates: Local authority rates; the rateable value and poundage should be stated.

8 External work

Repairs, maintenance and servicing in connection with external areas relating solely to the building being analysed should be recorded. Where external works relate to several buildings a separate analysis shall be prepared of this expenditure as it would be inappropriate to allocate costs to any one building. Sub-elements are as follows:

8·1 Repairs and decoration

8·2 External services: (work on drainage, water, electric or gas amenities).

8·3 Cleaning

8·4 Gardening

(ii) Perhaps one of the major invalidating factors affecting forecasting techniques is inflation. Throughout this book there are numerous graphs illustrating how building costs have risen during the short term. Figure 15 shows how, over the past 60 years or so, the tendency has been for building costs to rise. Two years ago clients were warned by building cost consultants that the rate of inflation on tender prices was 25%. Now, this incredible rate seems to have steadied, at least for a year or so, at around 12% per annum. Even so, that comparatively low inflation rate, when applied to buildings with an expected life of 80 years plus, produces figures that are beyond comprehension. One can perhaps imagine a client's reaction when being told to allow £20 per m² for renewing felt roofing in twenty years' time.

(iii) The client should have a working (if not detailed) knowledge of the techniques that are to be described shortly. When dealing with the larger companies this should be no problem.

(iv) In a world in which many of the natural resources are gradually disappearing for ever, the prices of various materials, fuels, etc., may not just be susceptible to regular inflation increases, but in addition to increases due to demand outstripping a quickly diminishing supply. For example, the energy crisis that was brought to everyone's attention in 1973, when oil and all oil-based products rose steeply in price, will have almost certainly successfully invalidated a good many cost forecasts. In this country, one result of the 1973 crisis was that

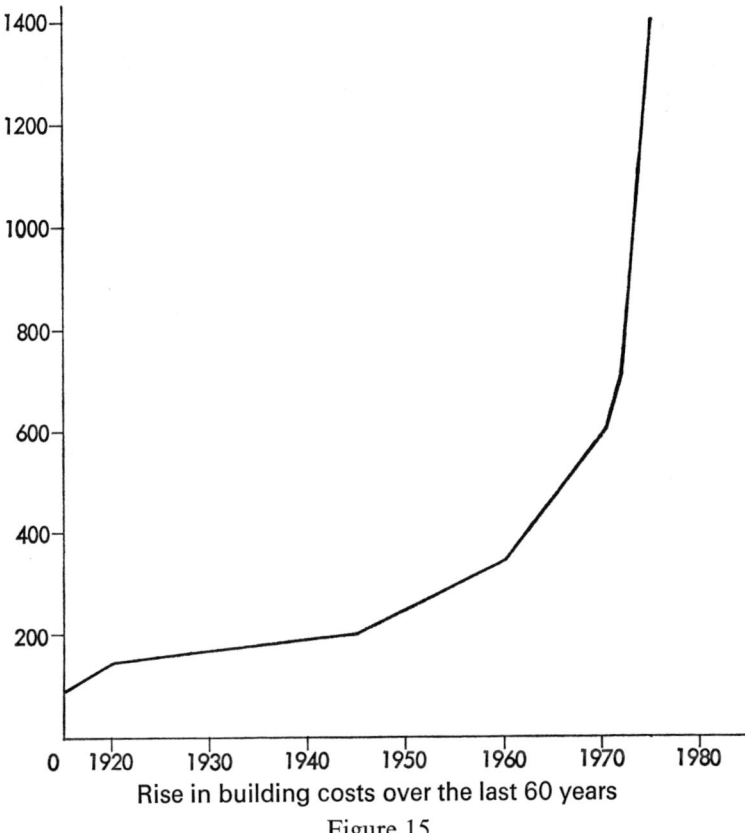

Rise in building costs over the last 60 years
Figure 15

the nationalized power industries, such as electricity, coal and gas, lost their heavy government subsidies. Electricity charges rose by 76% in 18 months, with similar increases in most other forms of energy and consequently the task of forecasting the most economic form of fuel for heating systems, faced with facts like these became very difficult.

However, despite these disadvantages the techniques described below are found to be of great use to clients, particularly the larger corporate bodies, but in the light of the comments above, they must be recognized to have limitations. With this in mind, it is good practice to present a client with a costs-in-use comparison as part of a report containing qualifications about inflation and the other factors that may affect the calculations.

There follows an example of a costs-in-use comparison for alternative building schemes, based on the information that was contained in Chapter 4. As well as studies like this one, the technique can also be used to compare individual elements, as shown in a later example.

Example
Calculate the costs-in-use for the three alternative building projects outlined below. The brief is to provide an office and a warehouse in the West Midlands for a commercial client. The design team has prepared three alternatives:

Scheme 1

Construction costs	£650000
Fees (say $12\frac{1}{2}\%$)	£80000
	£730000
Cost of site	£200000
Annual running costs	£12500
Replacement of finishes and services every 20 years	£30000
Ditto every 30 years	£20000

Scheme 2

Construction costs	£800000
Fees	£100000
	£900000
Cost of site	£150000
Annual running costs	£4000
Replacement of finishes and services every 20 years	£15000
Ditto every 40 years	£35000

Scheme 3
This alternative involves leasing a ready-built office block on the following terms:

Lease	50 years
Rent: premium	£40000
annual rent	£26000
fees (say 2%)	£1300
Annual running costs	£8000

When working out these examples it is assumed that the interest rate required is 8% and the client's rate of income tax is 35%. The expected life of the buildings in Schemes 1 and 2 is 50 years.

Scheme 1
Site
As the site is freehold, this first item, for the purposes of calculating an annual equivalent is taken to be in perpetuity, the rate of interest quoted is 8%, therefore the annual equivalent will be:
$8/100 = 0.08$ £200000 × 0.08 = £16000

Building costs
The total building costs and fees amount to £730000. As the building is expected to have a life of 50 years then to the annual equivalent calculated above it will be necessary to add an annual sinking fund to replace the wasting capital; i.e. £730000, in 50 years' time. This figure, by reference to the valuation tables (see Appendix B), is found to be 0.0017. Therefore,

Annual Equivalent	0.08
ASF	0.0017
	0.0817

Replacement costs
In addition to the land, building costs and running costs, the other major item to be considered in this type of calculation is replacement of services or finishes throughout the expected life of the building. In the scheme under consideration here it has been estimated that in addition to annual running costs, major replacement of services and finishes will be required at year 20, 30 and 40. The present day cost of replacement or redecoration may be found by multiplying the estimated future cost of replacement (see my earlier comments about inflation), by the *present value* of £1 at the appropriate rate per cent, for the number of years before the money needs to be spent. The reasoning behind this is that it is possible to calculate the smaller sum that could be invested now, to earn compound interest and so amount to the higher sum by the time that it will be required. It is

then possible to calculate the annual equivalent for this smaller sum together with the costs of the building land, etc., as shown below.

Site costs (from above) £16000

Building costs £730000

Replacement of finishes and services

First replacement at 20 years

present value of £1 @ 8 % over 20 years = 0·216

£30000
× 0·216
———
£6480

Second replacement at 30 years.

present value of £1 @ 8 % over 30 years = 0·099

£20000
× 0·099
———
£1980

Third replacement at 40 years.

present value of £1 @ 8 % over 40 years = 0·046

£30000
× 0·046
———
£1380
———
£739840
———

This figure should now be multiplied by the appropriate annual equivalent including an annual sinking fund,

$$0.0817 \times £739\,840 = £60\,445$$

Add   land costs £16000

annual running costs £12000
———

Total costs-in-use £88945
———

This exercise now needs to be carried out for Scheme 2.

Site costs: £15 000 × 0·08 = £12000

Building costs £900000

Replacement of finishes and services
First replacement at 20 years

	£15000		
	× 0·216		
	————	£3240	

Second replacement at 40 years

	£15000		
	× 0·046		
	————	£690	
	£35000		
	× 0·046	£1610	
	————		
		£905540	
		————	

£905540 × 0·0817 =		£73983
Add running costs		£4000
		————
Total costs-in-use		£189983
		————

Scheme 3

The terms of the lease for this scheme require a premium to be paid of £40000 In 50 years, at the end of the lease, a similar lump sum will have to be found, therefore, this item is treated as a wasting asset.

Premium	£40000	
Fees	£1300	
	————	
£41300 × 0·0817		£3374

To this should now be added
  (a) annual rent   £26000
  (b) the running costs that are to be the
      responsibility of the tenant   £8000
                                                                        ————
                              Total costs-in-use   £37268
                                                                        ————

Scheme 1. £89000
   „   2. £90000
   „   3. £37000

Therefore, clearly the most beneficial solution for the client in this case is to rent Scheme 3. However, there are certain considerations to be taken into account. The most obvious is that the lease may be renewable every 7 years, and with the review will almost certainly come a rent increase. So, by year 40, the annual rent could quite easily have increased four-fold. Experience shows that many clients who opt for this alternative initially, decide to build for occupation by the time the rental is reviewed for only the first time. This, then leaves the choice between schemes 1 and 2, between which, we have seen from our calculations, there is very little financial difference. The final decision should be taken in the light of factors outlined earlier in the book, viz.

(a) client taxation levels
(b) proposed use
(c) inflation

It is good that this example does not provide us with a clear cut answer, because in practice things very seldom are as easy as that. In such calculations judgement and decision making should also be seasoned with other, less easily analysed, considerations. As well as being able to be used for comparing complete buildings, as in the previous example, this technique can also be successfully employed when comparing individual elements or components. The following is an example of such a comparison:

Lift installation

Compare the overall costs of the following alternative types of lift installation:

(a) Oildraulic	lift installation	£10 000·00
	bore holes	£4 000·00
	builder's work	£1 000·00
	basic annual maintenance	500·00
	annual running costs	50·00
(b) Electric traction drive	lift installation	£10 000·00
	builder's work	£1 000·00
	basic annual maintenance	500·00
	annual running costs	200·00

Alternative (a)

It is assumed that the life of the lift is 30 years and therefore the calculation is to determine whether or not the extra initial costs associated with the oildraulic type of lift are offset by the low running costs.

Initial costs

lift	£10000
bore holes	£4000
builder's work	£1000

£15000 × i + ASF	
£15000 × 0·089 =	£1335
Add basic maintenance costs	£500
Add annual running costs	£50
Costs-in-use	£1885

Alternative (b)

£11000 × i + ASF	
£11000 × 0·089 =	£979
Add annual maintenance costs	£500
Add annual running costs	£200
Costs-in-use	£1679

Therefore, despite quite a large difference in initial costs, the long term costs vary little. Oildraulic lifts require a motor room at lowest level served, being smaller and more flexible in location. This simplifies room construction and builder's work and the elimination of a top motor room reduces the building height and the absence of top drive gear reduces stresses within the structure. Oildraulic lift equipment costs are obviously higher than for traction lift, but reduced building costs almost counteract this difference and allowing for cheaper maintenance and running costs (traction type lift systems have large electrically driven motors) oildraulic lifts become somewhat cheaper in the long term.

These then are the basic principles of design evaluation techniques and as I mentioned earlier they may successfully be applied to most of the elements that were discussed in Chapters 1 and 2.

References and Sources

Value in Building, Editors: G. H. Hutton and A. D. G. Devonald

Housing and Construction Statistics, HMSO

The Finchampstead Project, Building Research Establishment

The Economics of Industrialized Building, D. Bishop, MC, AMICE, ARICS

The Housing Cost Yardstick, HMSO

Development and Construction and Similar Procedures, HMSO

Building Maintenance Cost Information Service, Royal Institution of Chartered Surveyors

Building Cost Information Service, Royal Institution of Chartered Surveyors

Trench Fill, Cement and Concrete Association

R. Seifert and Partners, Chartered Architects and Town Planners

Davis, Belfield and Everest, Chartered Quantity Surveyors

R. E. M. Hayward, BSC, ARICS, Senior Lecturer, Polytechnic of the South Bank

Building, Magazine

Cost Benefits from variety reduction in Standard Steel windows, Building Research Station

Appendix A

1.   Substructure
2.   Superstructure
2.A Frame
2.B Upper floors
2.C Roof
2.D Stairs
2.E External walls
2.F Windows and external doors
2.G Internal walls and partitions
2.H Internal doors
3.   Internal finishes
3.A Wall finishes
3.B Floor finishes
3.C Ceiling finishes
4.   Fittings and furnishings
5.   Services
5.A Sanitary appliances
5.B Services equipment
5.C Disposal installations
5.D Water installations
5.E Heat source
5.F Space heating and air treatment
5.G Ventilating system
5.H Electrical installations
5.I Gas installations
5.J Lift and conveyor installations
5.K Protective installations
5.L Communication installations

Appendix B

Present Value of £1

No. of years (n)	Present worth of £1 payable in n years' time	
	(at 5%)	(at 10%)
5	0·78352	0·62092
10	0·61391	0·38554
20	0·37688	0·14864
40	0·14204	0·02209
80	0·02017	0·00048

Amount of £1

No. of years (n)	Amount of £1	
	(at 5%)	(at 10%)
5	1·2762	1·6105
10	1·6288	2·5937
20	2·6532	6·7275
40	7·0400	45·2593
80	49·5614	2048·4002

Annual Sinking Fund

No. of years (n)	Annual sinking fund	
	(at 5%)	(at 10%)
5	0·18098	0·16380
10	0·07951	0·06275
20	0·03024	0·01746
40	0·00838	0·00226
80	0·00103	0·00004

Appendix C

C1/SfB			
32			
Office block – 8			

BUILDING TYPE: Office Block	**OWNER/OCCUPIER:**
REGION: North West England	**DATE OF ERECTION:** August 1964

UPPER MANAGEMENT CRITERIA

The building is maintained to ensure the health and welfare of the staff employed in order to preserve a high level of efficiency in dealing with members of the public and to realise the building's economic life, estimated to be 60 years.

BUDGET PROCEDURE

Estimate: Estimates annually for maintenance jobs over £250, remainder from bulk estimate covering all buildings in depot estate and based on expenditure in the previous years.

Budget: Annual provision of funds based on estimates.

Cost control: Yearly computer printout of expenditure on the building. Monthly review to reconcile expenditure and commitments with Budget.

MAINTENANCE MANAGEMENT AND OPERATION

Responsibility: Maintenance Depot Superintendent
Total Estate: Maintenance Depot Superintendent
Routine Inspections: Yearly
Painting frequencies: Externally every 4 years. Internally as required
Cost records and feedback: Annual Computer Printout of Maintenance Cost on each coded building

Trades: ELECTRICIAN   FITTER
Number: As and when required

Incentive schemes: Productivity payment
Where directly employed labour force is used 80% is added to basic labour rates and includes cost of materials ex store
Work done by DEL 5% and contracted out 95%
Forms of contract: Measured Term Contract and Competitive Tender and Daywork Term Contract
Contract supervision: By Depot Technical Staff

BUILDING FUNCTION

Space use: Department Offices (Office Accommodation 88·5%) (Laboratory 4·0%) (Staff facilities 7·5%)
Number of occupants: 225
Design criteria: Central Administration. Standard Capital Cost Limit.
Change of use: N/A

Section D
Cost Analyse
March 1974

C1/SfB			
32 |	|	|	
Office block – 8			

BUILDING TYPE: Office Block	**OWNER/OCCUPIER:**
REGION: North West England	**DATE OF ERECTION:** August 1964

FORM OF CONSTRUCTION

Structure:	Concrete frame
External walls:	Solid wall 355 mm thick, faced sandstone externally plastered block-work internally. Cavity wall 280 mm thick, brickwork plastered internally. Solid wall 280 mm thick, RC panels externally, plastered brickwork internally.
Windows:	Painted galvanized steel casements
Roof:	127 mm RC slab, 25 mm screed, 3 layers felt and chippings
Internal partitions:	Dry partitioning 63 mm thick
Floor structure:	Solid concrete
Floor finishes:	Terrazzo to entrance hall and toilets. Lino to General Offices and corridors
Fittings & fixtures:	Teak laboratory worktops. Softwood counters in Public Offices
Internal decoration:	Walls and ceilings emulsion paint
External decoration:	Oil paint to windows and metalwork
Plumbing:	Copper water services and waste pipes. Cast iron internal drains and concrete cased.
Heating & ventilating:	Oil fired boilers LPHW
Lifts & Escalators:	2–5 persons (750 lb) lifts – A. & P. Steven Ltd, Serving 6 floors, 250 FPM
Other M & E Services:	Electric lighting and power; gas fired canteen/kitchen equipment; fire, intruder, lightning protection systems

PARAMETERS

Gross floor area	3748 m²	Height to ridge	— m	Storeys above (and	
Area of pitched		Height to eaves 'A'	19·8 m	including ground	
roofs (on plan)	Nil	(6 storey)		floor	6 No.
Area of flat roofs		Top of parapet 'B'	6·4 m	Floors below ground	
(on plan)	731 m²	(2 storey)		floor	—
Area of external		Area of external		Floor to ceiling	
glazing	900 m²	works	2200 m²	height	2·9 m

COMMENTS

FINANCIAL STATEMENT FOR YEAR 1971/72

Gross floor area: 3748 m²

C1/SfB
32 | | |
Office block 8

Element	Total £	Cost per 100 m² floor area	Brief description of work
0. Improvements & adaptations	£700·00	£18·68	Alterations to office accommodation
1. Decoration			
1.1 External decoration	—	—	
1.2 Internal decoration	—	—	
Sub-total	£ —	£ —	
2. Fabric			
2.1 External walls	—	—	
2.2 Roofs	80	2·13	
2.3 Other structural items	300	8·00	Doors, ironmongery and partitions
2·4 Fittings & fixtures	150	4·00	Notice boards, shelving and work-tops
2.5 Internal finishes	50	1·33	Plaster and lining board
Sub-total	£580	£15·46	
3. Services			
3.1 Plumbing & internal drainage	120	3·20	
3.2 Heating & ventilating	546	14·57	
3.3 Lifts & escalators	253	6·75	
3.4 Electric power & lighting	443	11·82	
3.5 Other M & E surfaces	59	1·57	Refrigerators, kitchen equipment (gas), lightning protection
Sub-total	£1421	£ 37·91	
4. Cleaning			
4.1 Windows	360	9·60	
4.2 External surfaces	—	—	Bi-monthly both sides
4.3 Internal	N/A	N/A	Costs not available
Sub-total	£ 360	£ 9·60	

FINANCIAL STATEMENT FOR YEAR 1971/72

Gross floor area: 3748 m²

C1/SfB
32
Office block 8

Element	Total £	Cost per 100 m² floor area	Brief description of work
5. Utilities			
5.1 Gas	192	5·12	
5.2 Electricity	1176	31·38	
5.3 Fuel oil	1695	45·22	
5.4 Solid fuel	—	—	
5.5 Water rates	283	7·55	
5.6 Effluents & drainage charges	—	—	
Sub-total	£3346	£ 89·27	
6. Administrative costs			
6.1 Services attendants			
6.2 Laundry			
6.3 Porterage	N/A		Cost not available
6.4 Security			
6.5 Rubbish disposal			
6.6 Property management			
Sub-total	£	£	
7. Overheads			
7.1 Property insurance	N/A		
7.2 Rates			
Sub-total	£	£	
TOTAL	£5707	£152·24	

External area 2200 m²	External works Total £	Cost per 100 m² of external area	Brief description of work
8. External works			
8.1 Repairs & decoration	—	—	
8.2 External services	100	4·54	Drain clearing
8.3 Cleaning	150	6·82	Cleaning paved areas
8.4 Gardening	90	4·09	Contract grass cutting
External Works Total	£ 340	£ 15·45	

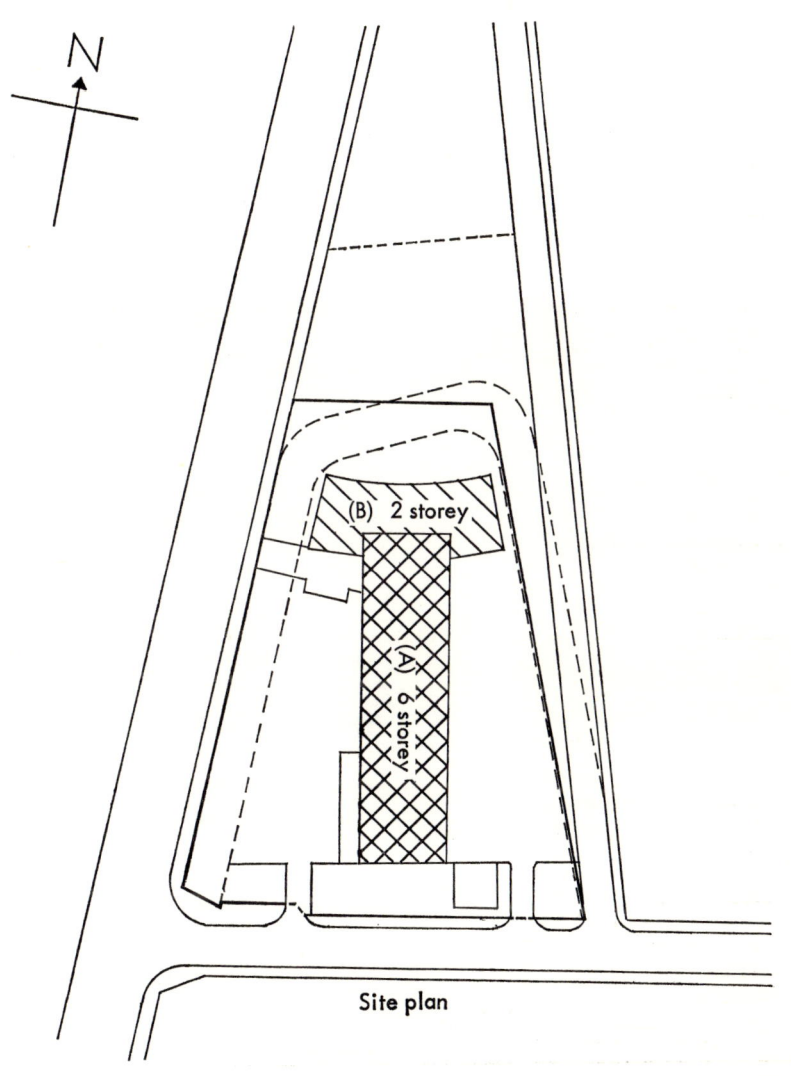

N

(B) 2 storey

(A) 6 storey

Site plan

Index